MARKET-BASED APPROACHES to ENVIRONMENTAL POLICY

MARKET-BASED APPROACHES to ENVIRONMENTAL POLICY

Regulatory Innovations to the Fore

Edited by

Richard F. Kosobud
Jennifer M. Zimmerman

VAN NOSTRAND REINHOLD

I(T)P®   A Division of International Thomson Publishing Inc.

New York · Albany · Bonn · Boston · Detroit · London · Madrid · Melbourne
Mexico City · Paris · San Francisco · Singapore · Tokyo · Toronto

I(T)P® An International Thomson Publishing Company
The ITP logo is a registered trademark used herein under license

Printed in the United States of America

For more information, contact:

Van Nostrand Reinhold
115 Fifth Avenue
New York, NY 10003

Chapman & Hall
2–6 Boundary Row
London
SE1 8HN
United Kingdom

Thomas Nelson Australia
102 Dodds Street
South Melbourne, 3205
Victoria, Australia

Nelson Canada
1120 Birchmount Road
Scarborough, Ontario
Canada M1K 5G4

Chapman & Hall GmbH
Pappelallee 3
69469 Weinheim
Germany

International Thomson Publishing Asia
221 Henderson Road #05–10
Henderson Building
Singapore 0315

International Thomson Publishing Japan
Hirakawacho Kyowa Building, 3F
2-2-1 Hirakawacho
Chiyoda-ku, 102 Tokyo
Japan

International Thomson Editores
Seneca 53
Col. Polanco
11560 Mexico D.F. Mexico

1 2 3 4 5 6 7 8 9 10 BBR 01 00 99 98 97

Library of Congress Cataloging-in-Publication Data

Kosobud, Richard F.
    Market-based approaches to environmental policy : regulatory
innovations to the fore / Richard F. Kosobud, Jennifer M. Zimmerman.
        p.   cm.
    Includes bibliographical references and index.
    ISBN 0–442–02483–5
    1. Pollution—Economic aspects—United States.   2. Pollution—
Government policy—United States.   3. Environmental policy—
Economic aspects—United States.   I. Zimmerman, Jennifer M.
II. Title.
    HC110.P55K67   1997
    363.73'0973—dc21   96–45349
                                                                CIP

http://www.vnr.com
product discounts • free email newsletters
software demos • online resources

email: info@vnr.com

a service of I(T)P®

CONTENTS

❑

PREFACE

❑ This volume consists of studies of regulatory innovation that originated as presentations to the Workshop on Market-Based Approaches to Environmental Policy in a series of meetings during the period 1992–1994. The studies were subsequently revised, taking advantage of the exchange of views that was an important feature of the Workshop format. This is not a book about the economic theory of regulatory reform and innovation, though that theory plays an important background role; rather, it is a book about new environmental control measures fresh from the drawing board or being drawn up and readied for consideration. It is also very much a book about the complex setting in which regulatory reform is debated, altered and, on occasion, implemented.

The Workshop was created as one way to explore this complex setting, drawing participants from the regulated, regulating, environmental, public interest, and academic communities. Participants brought separate and often contending viewpoints to the debate but appeared united in their willingness to consider, in a deliberative process, the strengths and limitations of new regulatory tools. At most meetings, formal discussants were chosen to comment on the study of the day. Discussants were chosen for their independent and critical views. Selected discussant comments are included in this volume. In addition, the open exchange of views that occurred in the Workshop was recorded and made available to authors for use in

ix

revisions of their work. This volume, therefore, can be said to be, in many respects, a joint Workshop product.

Regulatory reform has been very much in the air. In the environmental area, it has taken the form of a search for more flexible, cost-effective, and innovation-stimulating methods to achieve environmental goals. Such new methods could include improved direct regulation, reliance on voluntary efforts, or the creation of a wide variety of explicit incentives for households and producers to act in more environmentally benign ways. The major focus of the Workshop was on the latter—application of market-based incentives. These incentive techniques have a long history of mostly favorable academic analysis but a short history of use. The editors have selected studies of a subset of these market-based schemes that deal mainly with air pollution. There are two reasons for this choice: Air pollution ranks among our most serious problems, and air pollution has been the object of recent national legislation with significant implications for regulatory reform.

Market-based approaches are believed to have important strengths, in appropriate applications, compared with conventional regulation. These strengths have been, as yet, insufficiently tried and tested by their deployment, and their limitations have been insufficiently explored. The Workshop was initiated to help in this appraisal by providing a forum for discussing these strengths and limitations, for evaluating the first pieces of evidence on performance of those undergoing trial, and for deliberating on the design of measures under consideration.

During the period of Workshop meetings, environmental markets, or the use of tradable emission permits, moved to the center of the air pollution control stage and captured the attention of participants. A number of meetings were devoted to their design and implementation, especially the most innovative and comprehensive form of environmental market, the cap-and-trade model. So involved, contentious, and interesting did this discussion become that the Workshop organizers decided to try to record the various viewpoints by carrying out a sample survey of participants. Since this came at the close of the series of meetings on the topic, it could properly be termed a deliberative opinion poll of participant views (and not a random sample of the views of the population). The results of this poll are discussed at the close of Part 2.

The book is arranged in four parts. Part 1 provides an introduction and a new comprehensive survey of the current and potential cost savings that are accruing, or may accrue, from increased use of various types of market-based approaches. Part 2 contains studies of environmental markets under way or soon to be implemented and closes with the results of the deliberative opinion poll.

Part 3 ventures into the uncharted areas of possible future applications of new regulatory systems: one being a market for control of global warming and another exploring joint implementation prospects in this area. The next study in Part 3 examines the extended use of green taxes, and the last study describes a pioneering model for a more voluntary, decentralized environmental management. The incentive for the source of pollution in this instance is the avoidance of direct regulation by demonstrating that voluntary measures can produce, at the very least, an equivalent reduction in pollution. Part 4 contains a conclusion and conveniences for the reader, such as a glossary, index, and biographies of the contributors.

It is a pleasure to acknowledge the Workshop's support by a number of foundations and agencies. Generous financial support for Workshop activities was provided by the John D. and Catherine T. MacArthur Foundation. Further support came from the Chicago Council on Foreign Relations, the Institute of Government and Public Affairs at the University of Illinois, and the Department of Economics at the University of Illinois at Chicago. Facilities for meetings were arranged with the support of the U.S. Environmental Protection Agency, Region 5, Air and Radiation Division; the Commonwealth Edison Company; the Chicago Board of Trade; the Amoco Oil Company; and the Illinois Environmental Protection Agency.

The Workshop is itself the creation of many people of varied backgrounds and interests too numerous to single out; we are very happy to acknowledge their contributions in general. Foremost among this group are the participants: Where they have not clarified an idea during the deliberation, they have helped identify where the bone of contention lies. Others, not directly participating, supported the Workshop with suggestions. A number of people have helped more directly in the preparation of this manuscript. We should like to thank Professor Houston H. Stokes, Pamela Pinnow, Alex Mannella, Phil Nugyen, and Teresa Mieki. Adam Kosobud helped with some computer graphics.

The editors are much indebted to skilled and experienced publishing assistance from Van Nostrand Reinhold and owe a special debt to the cheerful and knowledgeable guidance of Jane Kinney, Senior Editor, Environmental Sciences. The editors retain sole responsibility for editorial viewpoints and for errors and omissions.

Richard F. Kosobud
Jennifer M. Zimmerman

❏

ACRONYMS

ACMA	Alternative Compliance Market Account
AERCO	Area Emissions Reduction Credit Organizations
ALAGC	American Lung Association of Greater Chicago
API	American Petroleum Institute
ATU	Allowance trading unit
BACT	Best available control technology
CAA'70.	Clean Air Act of 1970
CAAA'90	Clean Air Act Amendments of 1990
CAAPP	Clean Air Act Permit Program
CAC	Command-and-control regulation
CARB	California Air Resources Board
CBO	Congressional Budget Office
CBOT	Chicago Board of Trade
CCAP	Climate Change Action Plan
CEMS	Continuous emission monitoring systems
CERCLA.	Comprehensive Environmental Response, Compensation, and Liability Act of 1976
CFCs.	Chlorofluorocarbons
CO	Carbon monoxide
CO_2.	Carbon dioxide
EDF.	Environmental Defense Fund
EPRI	Electric Power Research Institute
ERC.	Emission reduction credits
ERMS.	Emissions Reduction Market System
FCCC.	Framework Convention for Climate Change
FERC.	Federal Energy Regulatory Commission

GCC.   Global climate change
GHG.   Greenhouse gas
HAP.   Hazardous air pollutant
ICC.   Illinois Commerce Commission
IEPA.   Illinois Environmental Protection Agency
IPCB.   Illinois Pollution Control Board
IPCC.   Intergovernmental Panel on Climate Change
JI.   Joint implementation
LAER.   Lowest achievable emission rate
LMOS.   Lake Michigan Ozone Study
MAC.   Marginal abatement costs
MACT.   Maximum achievable control technology
MSB.   Marginal social benefit
MSC.   Marginal social cost
MERC.   Mobile source emission credit
NAAQS.   National Ambient Air Quality Standard
NESCAUM.   Northeastern States Coordination Committee for Air
NO_x.   Nitrogen oxides
NPDES.   National Pollution Discharge Elimination System
NRDC.   Natural Resources Defense Council
NRRI.   National Regulatory Research Institute
NSR.   New Source Review
O_3.   Ozone
OTAG.   Ozone Transport Assessment Group
OTC.   Ozone Transport Commission
RACT.   Reasonably available control technology
RCRA.   Resource Conservation and Recovery Act
RECLAIM.   Regional Clean Air Incentives Market
SCAQMD.   South Coast Air Quality Management District
SIP.   State implementation plan
SO_x.   Sulfur oxides
TRI.   Toxic Release Inventory
U.S. EPA.   U.S. Environmental Protection Agency
U.S. IJI.   U.S. Initiative for Joint Implementation
ULEV.   Ultra low emitting vehicle
VMT.   Vehicle miles traveled
VOC.   Volatile organic compounds
VOM.   Volatile organic materials
WRI.   World Resources Institute
WWI.   World Wildlife Institute
ZEV.   Zero-emissions vehicle

1

❑

Market Tools

for Green Goals:

Regulatory Innovations

to the Fore

❑

1.1

❑

INTRODUCTION TO PART 1: REGULATORY REFORM AND REINVENTION

The Editors

❑ Although the air is cleaner, the water purer, and the land less contaminated in the United States than 25 years ago, we remain, by general agreement, short of our goals of protecting health and reducing economic and ecosystem impacts of pollution to acceptable levels. To reach those goals by further tightening the conventional regulatory or command-and-control measures threatens to increase marginal control costs, to limit control innovations, and to provoke nonproductive confrontation between regulated and regulating communities. Given these negative outcomes, it is easy to see that the decision, when choosing among alternative methods to achieve our environmental goals, is an important and sometimes difficult one.

Among the alternatives to conventional regulation, and the focus of these studies, is the expanded use of market-based approaches which, by reputation, can decentralize control decisions and create appropriate incentives for least-cost environmental management. While more often proposed than installed in the past, these incentive-type systems have received increasing attention by cost-conscious policymakers, have been applied in several instances, and have been considered for many more possible applications. An increasing pool of detailed designs, as well as a small but growing body of evidence on performance, are now available for evaluation. A major aim of this book is to contribute to the evaluation of the performance and design of deployed and potential incentive schemes.

To further this purpose, we selected studies by front-line researchers, administrators, and observers who are close to the institutional features and transactional processes that ultimately determine whether a market incentive proposal that is attractive in theory is successful in performance. These contributors came from the regulating and regulated communities, from environmental organizations, and from academia. Note that it is not a contention of this volume that the role of any of these essential communities in environmental affairs ought to be downgraded or emasculated. It is not that government ought to refrain from making key decisions in setting environmental goals or in establishing monitoring and enforcement procedures. Rather, it is a question of allocating to each community the subset of environmental decisions that it can best make in furthering our environmental ends.

In the endeavor to sort out these decisions, this volume is concerned mainly with problems of the concrete design and application of alternative environmental policy instruments and does not devote much space to the theoretical modeling of their comparative merits. However, we do not deny—and, in fact, consider as essential—the critical relationship between the two. Theoretical considerations can tell us whether there are worthy destinations in view. Implementation considerations tell us whether we can get there. It is important, and comforting, to know, for example, that the use of tradable pollution permits can be rigorously shown to result in a least-cost solution to control efforts.[1] It is equally important to recognize that wise application procedures or failure to recognize complications or departures from the abstract model can also affect the solution. These procedures and complications are brought to the fore in this volume.

Contributors do not neglect theoretical work; they are well aware of, and make reference to, a number of the findings. Economic theory affords a unifying framework for many of the studies of incentive system applications. Since providing such a framework was not a task assigned to any one study, and since it is likely to be convenient for some readers to have an account, however summary, of key ideas, the editors furnish in this introduction a quick survey of the core analytic arguments about the static and dynamic cost-effectiveness of the major alternative policy instruments.

---

[1]A technical account of the theory of instrument design can be found in the studies included in Part III of Dorfman and Dorfman (1993).

THE SIMPLE ECONOMICS OF LEAST-COST REGULATION

To examine why, in principle, we can expect savings by use of incentives compared with traditional regulatory systems, we set up a very simplified model of two cost-minimizing firms that differ in pollution control or abatement costs. That the firms aspire to minimize control costs and that they are two among many competitive emission sources are among the assumptions that simplify the exposition of this central case. The pollutants are not toxic or localized, and their impacts on households and other enterprises are externalities or social costs of production that have not yet been taken into account. Consequently, the government has set a target for their reduction but has not decided on the specific control measure to use.

To give a concrete feel for the range of these measures, it is useful to list examples of (1) direct, centralized or command-and-control (CAC), and (2) market-based or incentive techniques of control. The former include outright bans on harmful substances, uniform emission standards or specific technologies applied to all polluting sources regardless of individual control costs, content limits, and disposal requirements, among others. The latter include content and safety labeling, emission or product taxes, deposit refunds, disposal taxes, and various tradable permit or allowance schemes.

Figure 1.1.1, a–c graphs the two kinds of decisions important to our model: the government's choice of a target and policy instrument, and the firm's choices of control inputs (level of control). Both axes are scaled to the same spatial dimension for an easy grasp of the relationships.

The marginal abatement-cost curves (MAC) of the two firms, the incremental cleanup cost of emissions, are downward sloping as emissions increase. At the point of 1000 units, emissions are no longer reduced. The downward or negative slope assumes that less expensive control measures exist and are adopted first. The emissions of one firm do not interfere with the output and, hence, do not affect the emissions of the other.

In figure 1.1.1a, two firms, not necessarily of the same employment or output size or in the same industry, each emit 1000 units of pollutant in the no-control policy scenario. Firm 1 is less efficient at reducing emissions, as indicated by its MAC curve being above that of firm 2. Note that we are interpreting abatement costs to be limited to expenditures for control inputs or resources to secure compliance.

A CAC regulation carried out by specification of a particular technology or uniform emissions standard, is imposed by the government. Each firm is limited to the emission of 500 units. Firm 1

Cost-Effectiveness Comparisons

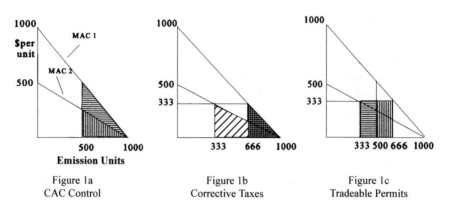

Figure 1a
CAC Control

Figure 1b
Corrective Taxes

Figure 1c
Tradeable Permits

Figure 1.1.1. Cost-effectiveness of three policy instruments compared.

reduces 500 units at a marginal cost of $500 for the last unit, and firm 2 reduces the same amount at a lower marginal cost of $250 for the last unit.

Society achieves a 50% reduction in emissions from the two firms — about the reduction in aggregate ozone precursor emissions required in many urban nonattainment areas and not far from aggregate sulfur dioxide emission reductions — at a total resource cost of the shaded triangles; that is, resources are drawn away from other uses to reduce emissions. Marginal control costs for the two firms are not equaled by application of this regulation, as is easily seen.

The government now turns to an incentive system that allows the firms to make specific control decisions in which they draw on their technical and economic knowledge. In figure 1.1.1b, a corrective tax of $333 per emission unit is levied by the government, which leads firm 1, via cost minimization, to reduce emissions by 333 units and to pay taxes on the 667 emitted units. Firm 2 reduces emissions by 667 units and pays taxes on the 333 emitted units. The corrective tax is set to achieve the same overall reduction in emissions as in the CAC case. No other tax level achieves the desired reduction, given the properties of these cost curves. Note that the government does not need information about the various control practices and techniques adopted, nor does the regulatory staff need to acquire and apply such knowledge. The government does need to obtain information about emissions, to monitor proper payment of taxes, and to develop appropriate enforcement procedures.

Cost-minimizing behavior on the part of each firm results in equal marginal costs of control of $333 per unit; that is, there is an incentive for the firm to look at each unit of reduction and to ask whether it can be controlled for less than the tax. When control costs mount above the tax, the firm will stop controlling and start paying the tax rate on the remaining emissions. Social resources allocated to control are measured in the shaded triangles of figure 1.1.1a and 1.1.1b. It is easy to prove, in this stylized situation, that the resources used in control are less under corrective taxes and tradable permits. Corrective taxes have internalized the social costs of pollution and have been cost-effective in the static sense in that a targeted amount of pollution has been efficiently reallocated among emitters.

Under the polluter-pays principle, firm 1 pays $333 per unit for the 667 emitted units, and firm 2 pays $333 per unit for the 333 units it emits. Firms may well argue — and many have — that this involves paying twice for emissions, once in the form of control measures and once in the form of taxes. Public interest groups may well counter — and many have — that the emissions that remain after the tax still impose harm on the community and, if the tax is optimal, that harm is measured by the tax rate. For our purposes, this issue may be set aside because tax revenues are transfers, not social costs; that is, tax revenues may be returned to the firms without undermining our environmental goal or they may be treated as a revenue source for other purposes.

Next, we remove all prior regulation and introduce an environmental market. In figure 1.1.1c, each firm is allocated 500 tradable emission permits free of charge. Permits may be banked, bought, or sold, but a permit must be given up for each unit of emission at the close of a given time interval.

If both firms simply turn their permits over to the government by emitting half and controlling half, we are back to the original costs of CAC. Providing that firms are allowed to talk to each other about trading permits, they will not be slow to realize that there are trading gains to be made. The low-cost firm can reduce emissions by more than half and offer to sell, to its advantage, surplus permits to the high-cost firm at a price at or above $250 (recall the length of side implications of the 30-degree angle of the right triangle). The high-cost firm can buy permits at a price at or below $500 and gain by controlling less and emitting more. To avoid problems of strategic behavior, assume that a trade is negotiated that neither party nor the government would want to change later. Or assume that, in a competitive market, the allocation of similar numbers of permits to firms of similar cost curves has led to an equilibrium price.

The optimum result would be that firm 1 buys 167 units from firm 2, or from the market to which firm 2 sells, at $333 per permit. Firm 1 reduces emissions by 333 and firm 2 by 667 units, the community's goal. The permit price is unique and could be determined by derivation of permit demand and supply curves for the market. We can say that the permit exchange is the best the firms can do for themselves. At the same time, it internalizes the social cost of pollution. No reallocation of 1000 units of emission reductions can be obtained at a lower compliance cost, given the permit price. The problem of environmental externalities resulting from market failures has been resolved by creating a market!

The outcome in the tradable permit example of figure 1.1.1c, with respect to each firms control of emissions and use of resources, is exactly the same as in the case of corrective taxes; that is, the control costs for firm 1's reduction of 333 units and the control costs for firm 2's reduction of 667 units are the same as they were in the corrective tax case. Decentralized decisions in the tradable permit market or in response to corrective tax rates, given our assumptions, lead to cost-effective results compared with centralized measures. This is the powerful core argument for environmental regulatory reform.

APPLICATIONS AND COMPLICATIONS

In the view of the editors, environmental regulation is an empirical science, and experience and data are required for trustworthy decisions on the selection of regulatory regimes. There are numerous problems to be resolved in implementing a regime attractive in theory but untested in practice; assumptions need to be examined, and difficulties treated lightly may turn out to weigh heavily in the final balance. The studies to follow will provide an account of the experience with, and the evidence on, these matters in particular areas of implementation. A brief survey here of the complications that must be added to the main results of figure 1.1.1 may help prepare the reader for the detailed discussions to follow.

Complications due to Changes in the Economy

Changes that occur in the economy could alter the static situation of figure 1.1.1. The advent of new control technologies may require the setting of new uniform standards or technology specifications under CAC regulation. While these adjustments due to technology events may appear to be simple in comparison with those required for changing incentive schemes, they often mean complex negotiation between regulated and regulating communities. Such regulation, along with the complications, could act to deter cost-saving innovation.

In the case of corrective taxes, the tax rate is a price determined by the government, and the amount of pollution reduction is a quantity determined by the enterprises. The incentive to innovate to avoid paying taxes is clear; any downward rotation or shift in the curves of figure 1.1.1 lowers the amount of taxes to be paid. Such innovations, if induced by the tax, lead to dynamic cost-effectiveness, a strong additional argument for incentive systems. However, changes up or down in the abatement-cost curves, affect the amount of pollution abated, so that tax rates need to be revised to secure the desired quantity. This may be painful politically, and frequent changes can introduce uncertainty into the regulatory process.

In the case of tradable permits, the amount of pollution reduction is a quantity determined by the number of permits issued by the government, and the permit price is determined in the market. Changes in the abatement-cost curves affect the price but not the quantity of pollution reduction, which is determined by the unchanged quantity of permits allocated. Again, there is an incentive for emitters to search for innovations that shift or rotate the cost curves downward—dynamic cost-effectiveness—as such shifts increase the number of permits that sources have available for sale or decrease the number that are advantageous to buy. Such innovations do not affect the aggregate quantity of permits issued unless the government makes a further determination based on revised costs.

If tax rates are set in money terms, fluctuations in the rate of inflation will change the cost curves and affect the quantity of pollution controlled, thus forcing the government to change the tax rate. In comparison, a permit market with well-informed traders results in permit price changes under inflation, not changes in the quantity of pollution reduced. The number of permits need not be altered.

At a market equilibrium for a given number of permits and, hence, a given amount of pollution control, the permit price provides a good estimate of current marginal control costs. If permits can be banked or are dated for future use, their current prices ought to be an estimate of discounted future control costs. In either case, risk management strategies can be developed explicitly through deriva-tive markets or through tailor-made transactions by intermediaries such as brokers.

Note that there are subtle differences between taxes and tradable permit schemes, beyond their shared decentralized attributes, that could affect their selection. One of these is revealed by considering the transfer amounts exchanged in the permit sale, as can be meas-ured in the shaded rectangles on either side of the 500-unit vertical line in figure 1.1.1c. Free allocation of tradable permits has left each firm better off in wealth than in the tax example (when tax revenues are not returned). This may be one interesting, and arguable, reason for selecting permits over taxes in that it could facilitate the cooper-ation of the regulated community in the transition to regulatory reform.

Complications due to Changes in Environmental Policy

The comparison among instruments takes another turn if and when a new target is established by environmental policy. In this situation, the CAC approach may seem both simpler and less fraught with uncertain consequences. The regulating community has to turn the technology valve up or down or write a new specification. However, the regulated community has the larger problem of adjustment. Changing tax rates presents problems for the government as dis-cussed earlier. Changing the volume of permits, by cancellation or discounting, introduces an element of uncertainty about the property right character of permits that may cause concern in the regulated community. A priori, it is difficult to weigh the balance of these considerations.

More complications arise in this comparison as more assump-tions of the simple models are relaxed. If we consider various transaction costs, lack of full and certain information among traders, departures from competitive price determination, and monitoring and enforcement costs, to mention only some of the problems, the results from our simple model become less straightforward and could even be reversed.

Only experience and empirical evidence on performance, in the light of these complications, can render a final judgment on the best regulatory measure for each individual application. In the studies that follow, the authors take pains to consider the factors that could affect the proper functioning of their respective incentive systems.

Although we cannot summarize their findings in a few sentences, at this introductory point, we can note that, in many, perhaps most, instances, cost savings remain significant and the main lessons of figure 1.1.1 continue to be valid despite the various complications.

WHY THE HARMS OF POLLUTION ARE NOT DISCUSSED MORE FULLY

Our attention has been focused on minimizing current or future control costs and not on minimizing the sum of both control costs and the harms caused by pollution. This has enabled us to concentrate on comparisons of alternate control mechanisms, given a targeted level of emission reduction, a considerable advantage in light of the many issues and choices to be examined in this area. It does mean that very important topics dealing with the harms or damages of pollution will not be discussed at any length and that problems of determining the optimal level of reduction of these harms will not be explored. In brief, we are engaged in neither an evaluation of the present targets of environmental policy nor a comprehensive benefit-cost analysis.[2]

The health, economic, ecosystem, and other impacts of pollution are increasingly being studied and estimated by the use of sophisticated tools such as clinical and epidemiological methodologies, contingent valuation, and hedonic indices, among others. The measurement or ranking of benefits of incremental improvements, long a difficult and contentious matter, has been advanced. These topics deserve a separate and full consideration, which is available in a number of publications.[3]

The workable alternative to comprehensive benefit-cost analysis, and the option chosen here, is to accept a target level of pollution reduction determined by a political process, such as characterizes much U.S. environmental legislation. The government having set the target, the emitters are free to turn to the least-cost measures for the attainment of that target. Consequently, the studies in this volume

[2]We confess to being supporters of increased use of benefit-cost analysis in appropriate situations. For a readable booklet containing comments by distinguished environmental economists and references to a burgeoning literature on the appropriate use of this tool, see Hahn and Portney (1996).

[3]A thorough and, to the uninitiated, demanding survey of the current state of benefit-cost analysis is available in Freeman (1993).

take as fixed the benefits to be achieved by that target.[4] This book is neutral on the question of the appropriateness of the targets determined by current policy. A very serviceable division of labor is achieved by setting to one side the matter of the target or ends for separate treatment from the selection of the appropriate measures or means.

A remaining question concerns the distributional impacts of regulatory improvements. To the extent that resources can be saved by more cost-effective control measures, they become available for alternative uses such as improving the welfare of low-income groups. There is no question, however, that taxes and tradable permits will affect the prices of final commodities that may occupy varying shares in the budgets of different income groups. Command-and-control measures have price implications as well. These distributional consequences are beyond the scope of this book and are best studied by a more complete analysis of benefits and costs, utilizing such specialized tools as computable general equilibrium models, which can take into account both direct and indirect price effects.[5]

HOW SIGNIFICANT ARE THE COST SAVINGS FROM THE USE OF INCENTIVES?

In our first study, Anderson, Carlin, McGartland, and Weinberger take on the critical but difficult task of putting a quantitative framework around the savings that are, or could be, realized by current or increased use of incentive systems. Their objective is to provide estimates for a comprehensive list of pollutants controlled by current policy. In carrying out this aim, they draw on and cite many currently available studies.[6] However, research on control costs is by no means complete and, in many instances, is not definitive. To extend the coverage, both for current and potential areas of application of incentive systems, the authors make use of their own experi-

---

[4]This position is analyzed in chapter 8 of Baumol and Oates (1988). This work also contains a survey of the present state of environmental theory and citations to the literature.

[5]A review of research in the area of environmental protection and its relation to impacts on different income groups may be found in chapter 15 of Baumol and Oates (1988).

[6]T. Tietenberg, 1985, has carried out an earlier survey of the cost-effectiveness of incentive systems, with an emphasis on emissions trading.

ence and judgment. Their position at the U.S. EPA gives them an excellent vantage point in this respect.

Only the enterprise's direct compliance expenditures under alternate regulatory regimes are estimated. These expenditures are only part of the welfare losses attributable to regulation. Other losses not included are private and public administrative and transaction expenditures (e.g., expenditures for database building and maintenance, for search and negotiation in the case of tradable permits, for monitoring and enforcement of rules, and for the collection of taxes). Also excluded are indirect losses due to general equilibrium (price effects), and growth effects. Estimating this more complete measure in order to compare regulatory regimes would require much more extensive data than are currently available and a much more complex modeling of welfare losses. Comparing the more limited measure of compliance expenditures for CAC and market-based approaches yields a useful first approximation. Furthermore, by comparing ratios of expenditures, the limitations of the concept may be offset, provided that nonmeasured expenditures occur in the same proportion as measured ones.

The study projects annual compliance cost savings for the year 2000 to be about \$11 billion (adjusted for inflation by using 1986 dollars) attributable to existing incentive mechanisms. An additional \$26 billion could be realized, it is estimated, with expanded use of incentive mechanisms. Consequently, savings of some 25% of total compliance expenditures at that future date are possible by the use of incentives. These savings result largely from the reallocation of pollution reductions to those sources able to carry them out more cheaply with existing control practices and technologies. Incentive systems also encourage research into the development of cheaper future control devices and practices, and the authors point out that the savings resulting therefrom could be even more significant. Although the quantitative significance of these future innovations remains unknown, it is possible to determine if companies are being stimulated to innovate by market-based incentives. The indicator would be research and development expenditures, the preliminary steps to technological innovation. An increase in expenditures under the market-based system would indicate a higher chance of future savings than there had been under CAC. Testing for this effect should rate high on the future research agenda.

The first study sets the stage for the other studies of this volume by finding that regulatory reform could reduce future compliance costs by 25%. Savings would, in all probability, be larger with an expanded concept of costs and would undoubtedly be larger if a good measure of technological progress were available. These savings are not automatic, of course, but rely on wise application procedures in particular instances. For this analysis, we turn to the studies in later parts.

REFERENCES

Baumol, William, J., and Wallace C. Oates. 1988. *The Theory of Environmental Policy*. Cambridge, U.K.: Cambridge University Press.

Dorfman, Robert, and Nancy S. Dorfman (eds.) 1993. *Economics of the Environment*. 3rd ed. New York: W. W. Norton & Company.

Freeman, Myrick A. III. 1993. *The Measurement of Environmental and Natural Resource Values: Theory and Methods*. Washington, DC: Resources for the Future.

Hahn, Robert W., and Paul R. Portney. 1996. *Benefit-Cost Analysis in Environmental, Health, and Safety Regulation: A Statement of Principle*. La Vergne, TN: Publisher Resources, Inc.

Portney, Paul R. 1994. "Economics of the Clean Air Act," *Journal of Economic Perspectives* 4 (fall):173–181.

Tietenberg, T. H. 1985. *Emissions Trading: An Exercise in Reforming Pollution Policy*. Washington, DC: Resources for the Future.

1.2

❑

COST SAVINGS FROM THE USE OF MARKET INCENTIVES FOR POLLUTION CONTROL

Robert C. Anderson, Alan Carlin, Albert M. McGartland,

and Jennifer B. Weinberger

❑ I have suggested that for many social policies, and for a pollution policy in particular, there are three main strategies of implementation: regulation; subsidization; and some system of charging for user rights. My own thinking on these matters leads me strongly to favour the charging system wherever practicable.

J. H. Dales (1968, p. 105)

Market incentives should be used to achieve environmental goals, whenever appropriate.

President Bill Clinton and Vice President Al Gore. (1995, p. 6)

Although the United States has historically relied heavily on command-and-control (CAC) regulations to control pollution, economists and public officials have long emphasized the advantages of market incentive approaches to environmental pollution control over traditional regulatory schemes. Market-based approaches to pollution control provide an opportunity for achieving better results from environmental spending in terms of both cost savings and increased innovation. The U.S. Environmental Protection Agency (EPA) (1990) estimates that the total annualized cost of all types of pollution control will reach $125 billion in 1995 (2.4% of GNP) and will increase to $148 billion by the year 2000 (2.6% of GNP), assuming implementation of existing and proposed regulations. The potential for cost savings from comprehensive use of economic incentives could be significant.

Both the President and the 104th Congress emphasized the promise and importance of market incentives in reaching the country's environmental goals. The heavy emphasis on regulatory reform in Congress and the executive branch makes it an opportune time to reexamine the potential for cost savings and the opportunities for increased innovation resulting from market-based approaches to pollution control. While other studies have examined cost savings achieved with market incentives in specific applications, no one has assessed the cost savings from applying market incentives across environmental programs. This study provides such broad examination and aims to illustrate very roughly the potential savings in order to invoke further debate over their widespread use. The first section provides rough estimates of the cost savings from existing EPA market incentive programs. The next section provides some admittedly crude estimates of potential cost savings if market incentives are adopted in other programs as well. The final section offers a brief discussion of the opportunities for increased innovation under market incentive pollution control.

DEFINITION OF MARKET INCENTIVES

Market incentives are defined here as instruments that aim to in-duce—rather than command—changes in behavior by providing financial or other motivations for sources to reduce releases of pollution. Traditional CAC approaches to pollution control do not provide incentives for polluters to reduce the quantity or rate of releases below allowable limits or to improve the quality of releases of pollutants beyond that allowed by regulation. Examples of econ-omic incentives include pollution taxes or charges, fees, tradable discharge permits or pollution allowances, and community right-to-know provision programs.[1] The definition used here does not include mechanisms that price activities having pollution as a by-product, such as high-occupancy vehicle lanes or parking surcharges, which do not place an explicit or implicit price on incremental units of pollution.

---

[1]The annual Toxics Release Inventory, detailing toxic releases for each major manufacturing facility, is available to the public. It is considered a market incentive because it provides an incentive for polluters to reduce their toxic releases for public relations and other reasons.

APPROACH TO QUANTIFYING COSTS

In a few cases examined here, published studies provide estimates of the cost savings resulting from particular economic incentive programs. Where no published estimate exists, we first ascertain whether the market-based program is likely to produce significant compliance cost savings. If so, we make a rough estimate of the likely cost savings by reducing unit production cost or pollution control cost figures by an estimated percentage.

Most of our estimates of unit cost savings come from the literature examining incentive systems. The authors find savings in control costs and/or improvements in environmental quality relative to a CAC system. Theoretical modeling and simulation of market incentive and CAC systems consistently demonstrate that incentive systems outperform other regulatory approaches in terms of efficiency. Table 1.2.1 summarizes many of the quantitative studies performed by economists. The final column of this table presents a ratio of CAC costs to the least-cost approach to meeting the same objective using economic incentives. A ratio of 1.0 suggests that the CAC approach is equal (in terms of costs) to the economic incentive approach, and so the savings are zero. A ratio greater than 1.0 means that there are positive potential savings from using economic incentives. Realized cost savings, however, may often fall well short of projections. Trades are fewer and cost savings smaller than indicated by economic modeling. A number of explanations have been offered for why the full savings have not always been realized; most point to the failure of these models to adequately reflect regulatory, institutional, transaction (information), and legal constraints (Atkinson and Tietenberg 1991, Dudek and Palmisano 1988, Hahn 1989, Hahn and Hester 1989, and Liroff 1986). Nonetheless, even if the cost savings are less than predicted, the actual savings are still impressive, especially when the stronger stimulus provided by market incentive approaches for innovation and technical change is taken into account.

The costs presented in this chapter represent estimates of direct regulatory implementation and compliance costs.[2] Transaction costs — such as those associated with searching for, or attempting to sell, pollution credits or allowances; costs of negotiating a transaction; and monitoring or litigation-related costs — are not considered.

[2]Note that the costs considered here do not reflect the social costs of pollution control regulation as they do not attempt to approximate the opportunity cost of goods and services forgone as a result of resources diverted to environmental protection.

Table 1.2.1. Quantitative Studies of Economic Incentive Savings

Pollutants controlled	Study, year and source	Geographic area	Command-and-control (CAC) approach	Ratio of CAC cost to least cost
Air				
Criteria air pollutants				
Hydrocarbons	Maloney and Yandle (1984), T	Domestic DuPont plants	Uniform percentage reduction	4.15[a]
Lead in gasoline	U.S. EPA (1985), A	United States	Uniform standard for lead in gasoline	See footnote for $ savings[b]
Nitrogen dioxide (NO$_2$)	Seskin et al. (1983), T	Chicago	Proposed RACT regulations	14.4
NO$_2$	Krupnick (1986), O	Baltimore	Proposed RACT regulations	5.9
Particulates (TSP)	Atkinson and Lewis (1974), T	St. Louis	SIP regulations	6.0[c]
TSP	McGartland (1984), T	Baltimore	SIP regulations	4.18
TSP	Spofford (1984), T	Lower Delaware Valley	Uniform percentage reduction	22.0
TSP	Oates et al. (1989), O	Baltimore	Equal proportional treatment	4.0 at 90 μg/m^3
Reactive organic gases/NO$_2$	SCAQMD (spring 1992), O	Southern California	Best available control technology	1.5 in 1994
Sulfur dioxide	Roach et al. (1981), T	Four Corners area	SIP regulation	4.25
Sulfur dioxide	Atkinson (1983a and b), A	Cleveland		About 1.5

Table 1.2.1. Continued

Pollutants controlled	Study, year and source	Geographic area	Command-and-control (CAC) approach	Ratio of CAC cost to least cost
Sulfur dioxide	Spofford (1984), *T*	Lower Delaware Valley	Uniform percentage reduction	1.78
Sulfur dioxide	ICF Resources (1989), *O*	United States	Uniform emission limit	5.0
Sulfates	Hahn and Noll (1982), *T*	Los Angeles	California emission standards	1.07[d]
Six air pollutants	Kohn (1978), *A*	St. Louis		
Other				
Benzene	Nichols et al. (1983), *A*	United States		
Chloroflouro-carbons	Palmer et al. (1980), Shapiro and Warhit (1983), *T*	United States	Proposed emission standards	1.96
Airport noise	Harrison (1983), *T*	United States	Mandatory retrofit	1.72[e]
Water				
Biochemical oxygen demand (BOD)	Johnson (1967), *T*	Delaware Estuary	Equal proportional treatment	3.13 at 2 mg/1 DO; 1.62 at 3 mg/1; 1.43 at 4 mg/1
BOD	O'Neil (1980), *T*	Lower Fox River, Wisconsin	Equal Proportional treatment	2.29 at 2 mg/1 DO; 1.71 at 4 mg/1; 1.45 at 6.2 mg/1

Table 1.2.1. Continued

Pollutants controlled	Study, year and source	Geographic area	Command-and-control (CAC) approach	Ratio of CAC cost to least cost
BOD	Eheart et al. (1983), *T*	Willamette River, or...	Equal proportional treatment	1.12 at 4.8 mg/1; 1.19 at 7.5 mg/1
BOD	Eheart et al. (1983), *T*	...Delaware Estuary in PA, DE, and NJ	Equal proportional treatment	3.00 at 3 mg/1 DO; 2.92 at 3.6 mg/1
BOD	Eheart et al. (1983), *T*	Upper Hudson River in NY State	Equal proportional treatment	1.54 at 5.1 mg/1; 1.62 at 5.9 mg/1
BOD	Eheart et al. (1983), *T*	Mohawk River in NY State	Equal proportional treatment	1.22 at 6.8 mg/1
Heavy metals	Opaluch and Kashmanian (1985), *O*	Rhode Island jewelry industry	Technology-based standards	1.8
Phosphorous	David et al. (1977), *A*	Lake Michigan		

[a]Based on 85% reduction of emissions from all sources.

[b]The trading of lead credits reduced the cost to refiners of the lead phasedown by about $225 million.

[c]Ratio based on 40 g/m³ at worst receptor, as given in Tietenberg (1985), Table 4.

[d]Ratio based on a short-term, 1-hr average of 250 g/m³.

[e]Because it is a benefit-cost study instead of a cost-effectiveness study, the Harrison comparison of the CAC approach with the least-cost allocation involves different benefit levels. Specifically, the benefit levels associated with the least-cost allocation are only 82% of those associated with the CAC allocation. To produce cost estimates based on more comparable benefits, as a first approximation, the least-cost allocation was divided by 0.82 and the resulting number compared to the CAC cost.

Acronyms used: CAC: command-and-control, the traditional regulatory approach. DO: Dissolved oxygen; higher DO targets indicate higher water quality. RACT: reasonably available control technologies. SIP: state implementation plans.

Sources: *A* stands for Anderson et al. (1990); they did not compute the ratio or provide the other information left blank in this table. *O* stands for original reference. *T* stands for Tietenberg (1985), Table 5. See Bibliography for all references.

The estimates of savings potential presented here are conservative to the extent that traditional CAC approaches are unable to duplicate the pollution control results achieved by incentives. For example, the incentives provided to households by volume-based solid waste disposal pricing might be difficult to achieve through direct regulatory approaches.

SUMMARY OF COST SAVINGS RESULTS

Table 1.2.2 summarizes the cost savings estimated in the following pages from the use of market incentives. The table puts the savings into perspective relative to the total national costs of pollution control in the air, water, and land media. The estimated savings from the use of market incentives are divided into three categories: savings from existing incentive mechanisms, projected savings from existing incentives in the year 2000, and *additional* possible savings in the year 2000 from expanded use of incentives. These estimated amounts are on an annual basis.

Projected savings from existing incentives are estimated to have been approximately $8 billion in 1992, and are projected to increase to over $11 billion by the year 2000. Additional potential savings in 2000 from expanded use of incentives are estimated to be about $26

Table 1.2.2. Summary of Estimated Cost Savings from use of Economic Incentives (in Billions of Dollars)

Media	Estimated environmental compliance costs[a] in 1992	Savings due to existing incentive mechanisms	Estimated environmental compliance costs[a] in 2000	Projected savings in 2000 from existing incentives	Additional possible savings in 2000 with expanded incentives
Air	28.9	1.7	43.5	2.8	12.6
Water	45.1	4.5	57.4	5.5	11.7
Land	32.6	1.9	45.6	3.0	2.1
Total	$106.6	$8.1	$146.5	$11.3	$26.4

[a]Environmental compliance costs exclude those costs of government environmental regulatory agencies that are related predominantly to promulgation and enforcement of regulation. These costs are not likely to be reduced through increased use of economic incentives. Estimates of compliance costs are in billions of 1986 dollars annualized at 7%.

Source: Estimated compliance cost estimates from U.S. EPA (1990). Estimated savings estimates are explained in the text that follows below.

billion. When the projected savings in 2000 from existing incentives are added to the possible savings in 2000 from expanded use of incentives, total potential savings would be nearly $38 billion, or 26% of estimated compliance costs in the year 2000.

While many studies of specific market incentive programs have found savings potentials significantly higher than 26% (as illustrated in Table 1.2.1), our estimates illustrate the magnitude of savings that might be achieved based on conservative assumptions and excluding numerous existing and expanded incentive programs for which cost savings estimates cannot be calculated.

COST SAVINGS FROM EXISTING INCENTIVE PROGRAMS

Air Pollution Control Programs

The U.S. EPA (1990) identifies a number of incentive systems that involve air pollution, including fee-based programs, several tradinq systems, and an information program, among others. In the following pages, we estimate the cost savings resulting from these incentive programs relative to a command-and-control approach. For two of the incentive programs, cross-line emissions averaging and chloro-fluorocarbon taxes, there is no basis for estimating the magnitude of cost savings expected. Another one of the programs, fire and wood stove permit trading, is expected to yield savings that are unlikely to have national significance. Consequently, these three programs will not be examined below. For the others, we attempt to assess system-atically the current and projected savings from each of these systems relative to CAC approaches.

Air Emissions Trading

The EPA's air emissions trading program consists of four separate activities: bubbles, offsets, banking, and netting. The components of the air emission trading program were developed through regulations and policy statements issued by EPA. The programs began indepen-dently in the mid- to late 1970s and culminated in EPA's Final Emissions Trading Policy Statement, dated December 4, 1986.

Hahn and Hester (1989) estimate that bubbles produced compli-ance cost savings of $435 million over six years (about $70 million per year), offsets yielded negligible savings, banking resulted in very small savings, and netting saved some $525 million to $12 billion over six years, or $90 million to $2 billion per year.

Federal Nonattainment Area Fees

The 1990 Clean Air Act Amendments provide for a fee of $5,000 per ton on "excess" emissions of volatile organic compounds in certain ozone nonattainment areas. Excess emissions are emissions that exceed 80% of baseline quantities. The fees become applicable at varying dates 15 to 20 years after passage of the act, depending on the severity of an area's ozone nonattainment problem. At $5,000 per ton, such fees equal or exceed incremental control costs in all but the Los Angeles area. Consequently, they could have a significant impact and should be more cost-effective than CAC alternatives. Any cost savings, however, would not occur until well in the future since the fees would be unlikely to affect compliance costs until the 15- to 20-year periods are nearing expiration.

Chlorofluorocarbon Allowance Trading

In 1988, the United States ratified the Montreal Protocol on Substances That Deplete the Ozone Layer. The Montreal Protocol called for a cap on production of chlorofluorocarbons (CFC) at 1986 levels, with a phaseout of production by the year 2000. Title VI of the Clean Air Act Amendments of 1990 calls for additional restrictions on CFC production. In late 1991, EPA issued a temporary final rule that apportions baseline allowances, provides a schedule for reducing the allowances, and allows for trading of allowances among firms (U.S. Environmental Protection Agency, 1991b and 1991c).

Nearly 600 million kg of CFCs are apportioned among producers in the EPA rule. Allowing trading among producers helps to assure that, as CFC production is phased out, production becomes concentrated at the most efficient facilities. By 1995, the wholesale price of most CFCs was in the range of $10 to $15 per pound, indicating that trading provisions potentially have a large value to producers.[3] However, CFCs are also subject to a windfall profits tax that began at $1.37 per pound in 1990, rose to $3.10 per pound by 1995, and increases at $0.45 per pound each year thereafter. In the context of windfall profits taxes, trading provisions have a value if the substances continue to be competitive in the marketplace (as they have to date). As production is phased out, trading is likely to become more valuable per pound. The value is difficult to establish; rather arbitrarily, we place it at an average of $1 per pound in 1992, rising to $2 per pound by 1996. This would indicate that trading provisions might have saved about $500 million annually in 1992 and a

[3]Personal communication with A. N. Other, DPont Corporation, October 17, 1995.

comparable amount on a lower volume in 1996. By the year 2000, CFC phaseout will be complete so that there would be no potential for savings from trading beyond that date.

Acid Rain Allowance Trading

Under authority of Title V of the Clean Air Act Amendments of 1990, the EPA established a program to reduce SO_2 emissions from electric utilities. Under the EPA rules, existing utility sources are granted allowances and may meet their emission limits by using their allowances or by acquiring allowances from utilities that control emissions beyond what is required.

Because it encourages cost-effective emissions control, the acid rain allowance trading system could save affected utilities substantial amounts relative to a CAC alternative. The EPA estimates of cost savings run from $700 million to $1.0 billion per year, beginning in the mid- to late 1990s (ICF, Inc. 1991).

Reclaim

In October 1990, California's South Coast Air Quality Management District (SCAQMD) began developing a new large-scale trading regime called the Regional Clean Air Incentives Market (RECLAIM). The program was adopted in October 1993 after three years of debate, public hearings, and rule revisions. The rules implementing the program became effective on January 1, 1994. The RECLAIM program initially involved 390 facilities, all major sources of nitrogen oxide emissions. In addition, 41 of these facilities constitute a market for sulfur oxide emissions.[4] The facilities in the program can choose from any method available to meet required emissions reductions, such as purchasing traded emissions, installing equipment to control pollution, or developing methods to prevent pollution.

The SCAQMD projects that the RECLAIM program would, in its early years, reduce compliance costs relative to the CAC alternative from $660 million to $223 million, for a saving of $437 million (1987 dollars). By 1997, the marketable permit program is projected to reduce compliance costs from $930 million to $636 million for a

---

[4]The original RECLAIM proposal also covered emissions of volatile organic compounds (VOCs); however, the inclusion of VOCs was postponed. After the program was approved in October 1993, SCAQMD staff began new efforts to expand the RECLAIM program to include VOC emissions. Since SCAQMD aims ultimately to include VOCs in the program, we include them in our estimates of cost savings associated with the program.

saving of $294 million (1987 dollars). In other words, the SCAQMD projects that RECLAIM would yield savings in compliance costs of about 65% in the early years of the program and about 32% in later years as pollution control requirements are tightened to levels that require controls on nearly every source and leave little choice as to control options.

Scrapping Vehicles

The EPA recently issued guidance for states that wish to scrap vehicles as a means of demonstrating continued progress in reducing emissions for their state implementation plans. The guidance covers pre-1981 vehicles, about 30 million of which are still in operation. Recent estimates show that scrapping older vehicles may reduce hydrocarbon emissions at a cost ranging from $3,000 to $7,000 per ton, a range comparable to what is currently spent for stationary source controls in many parts of the nation (Schroeer (1992), OTA (1992), Hahn (1993).

Certainly, only a fraction of the 30 million vehicles that could be candidates for scrapping are likely to be scrapped as part of such a program. To obtain a rough estimate of the potential savings from scrapping programs, we assume that 1% of the total candidate population is scrapped annually through a state scrappage program. We further assume that the cost-effectiveness of alternative controls is $10,000 to $12,000 per ton in areas where scrapping programs operate. This yields an average saving of at least $5,000 per ton. Finally, we assume that each vehicle scrapped will reduce emissions by approximately 0.13 ton. Thus, vehicle scrapping could reduce pollution by nearly 40,000 tons, and the potential annual pollution control cost savings for vehicle scrapping could be as much as $200 million if vehicle scrappage programs are adopted in most areas where they are cost-effective.

SARA Title III

Although SARA Title III is not typically thought of as an air program, it does provide incentives to reduce air emissions, as well as releases to water and land. SARA Title III, also known as the Emergency Planning and Community Right to Know Act, is widely viewed as more effective than direct regulation. Under SARA Title III, firms that manufacture or process over 25,000 pounds of some 302 chemicals (currently) must report all quantities released to the air or water, transferred to public sewage-treatment plants, deposited on land, or injected underground. These results are then made available to the public annually. Current control costs for several of these substances now range from less than $500 to more than $5,000 per ton.[5]

Firms are likely to identify inexpensive control measures to bring reported quantities down and to avoid expensive options. Because firms effectively have infinite choice as to which measures are adopted, the cost-effectiveness of options selected is likely to be quite good. Thus, it appears reasonable to assume that incremental control costs for releases to air eliminated as a consequence of SARA Title III would average $2,000 per ton ($1 per pound) less than CAC regulations that accomplished the same reductions.

The extent of emission reductions that one might expect from SARA Title III is somewhat difficult to project but can be roughly estimated based on recent reporting trends. Reportable air emissions declined from close to 1.9 to nearly 1.7 billion pounds from 1992 to 1993, a decline of 10.4%, saving an estimated $193 million in compliance costs. Between 1988 and 1993, reportable air emissions declined from 2.7 billion pounds in 1988 to 1.7 billion pounds in 1993, for an annual average decline of 7.8%. Projecting impacts to the year 2000 at the rate of 7.8% per year, one obtains a cumulative reduction of over 45% relative to 1993 levels. Assuming average savings of $1 per pound, annual savings by the year 2000 would be about $108 million.

Heavy-Duty Truck Emissions

Title III of the Clean Air Act (as amended) calls for an emissions standard for nitrogen oxides that represents the maximum degree of reduction achievable, with a goal of attaining a reduction of 75% in the "average of actually measured emissions" from heavy-duty gasoline engines. Implementation of these requirements by EPA allows manufacturers to comply by averaging together the emissions performance of all the heavy-duty engines they produce.

Compliance cost savings for engine manufacturers are unknown but could amount to $100 or more per engine. According to *MVMA Motor Vehicle Facts and Figures* (1994), diesel engines were installed in 85% of the trucks over 14,000 pounds gross vehicle weight (gvw). In recent years, domestic manufacturers produced about 130,000

---

[5]For example, the 1995 petroleum refinery NESHAPS rule controls 13 hazardous air pollutants (HAPs), all of which are listed under SARA Title III at an average cost of approximately $500 per ton.

heavy-duty trucks (33,000 pounds gvw and over) and 70,000 me-
dium-duty trucks (14,000–33,000 pounds gvw). Assuming that about
$100 is saved on each of 30,000 engines (15% of 200,000), the
averaging provisions for truck engines could save manufacturers
about $3 million annually. While the figure of $100 savings per
engine may not be reliable, the calculation illustrates that, even if
one assumes much higher average savings of $500 per engine, the
aggregate savings from this provision are unlikely to exceed $20
million annually.

Clean-Fuel Vehicle Credit Program

Section 249 of the Clean Air Act (as amended) requires that
automobile manufacturers offer for sale in California 150,000 clean-
fuel vehicles in model years 1996–1998 and 300,000 clean-fuel
vehicles in model year 1999 and thereafter. Emission characteristics
required for 1999 and subsequent model years are stringent — equal
to the low-emission vehicle (LEV) standard. Manufacturers will be
obligated to make clean-fuel vehicles available in proportion to their
California sales volume. A fully marketable credit program accom-
panies these requirements. As the *Federal Register* notice observes,
"By purchasing credits through trading, small volume manufacturers
would help defray the cost of developing the clean-fuel vehicle
technology..." (U.S. Environmental Protection Agency, 1991a).

Since there are likely to be significant economies of scale in the
production of clean-fuel vehicles, manufacturers holding a small
share of the California market are likely to find it less expensive to
acquire credits in the marketplace than to develop their own clean-
fuel vehicle. Since a clean-fuel vehicle is likely to cost from a few
hundred to perhaps $1,000 more to produce than an ordinary
vehicle, when these vehicles are produced in large volume, the
savings from trading could average on the order of $200 per vehicle
for model years 1996–1998 and $500 per vehicle for 1999 and later
model years, since inefficient, low-volume production would be
avoided. Thus, we estimate that the credit trading program for
low-emission vehicles could produce compliance cost savings of $30
million per year in the years 1996–1998 and $150 million per year
thereafter.

Oxygenated Gasoline Credit Program

The 1990 Clean Air Amendments require that gasoline with a
2.8% oxygen content be marketed in 39 cities during the winter

months beginning in 1992 and that reformulated gasoline with an oxygen content of 2% be marketed by 1995 during the summer months in nine cities. Additional cities may opt into the reformulated gasoline requirements. As part of this rule, the 1990 Amendments allow refiners and blenders to trade oxygen credits. This trading provision could save refiners up to $150 million per year, calculated as follows. The oxygenated fuel requirements apply to as much as 30 billion gallons of fuel annually. The cost of adding oxygen is approximately 3 cents per gallon; trading might save as much as 0.5 cent per gallon of this cost, or $150 million. The more modest goals of 1992 might have saved refiners as much as $20 million according to our estimates.

Summary of Cost Savings from Existing Incentives to Reduce Air Pollution

Air emissions trading apparently constitutes the largest source of incentive-based savings, currently accounting for an estimated $160 million to $2.1 billion in annual savings (midpoint $1.0 billion). The total current savings from existing incentives to reduce air pollution are estimated at $1.7 billion, all of which pertain to stationary sources.

By the year 2000, a number of additional incentive programs for stationary sources will begin to have an effect: acid rain trading will save an estimated $700 million to $1.0 billion annually; SARA Title III will save approximately $108 million annually; RECLAIM will save perhaps $300 million annually; and scrapping older vehicles will save roughly $200 million annually. Air emissions trading other than RECLAIM may produce future savings comparable to current levels, or $1.0 billion annually. The total cost savings for existing stationary source programs in the year 2000 are thus about $2.5 billion. Mobile source savings in the year 2000 include oxygenated fuels, heavy-duty truck engines, and clean-fuel vehicle trading programs; collectively, these provisions may produce cost savings of about $300 million annually.

Water Pollution Control Programs

Several incentive mechanisms are currently being used to influence either the cost or magnitude of water pollution activities (EPA 1990). These programs include effluent charge systems, reporting requirements, liability rules, and trading of effluent discharge credits between point and nonpoint sources or among point sources. Point source programs dominate this list; the only nonpoint incentive programs involve trading with point sources. For the liability rules under the Comprehensive Environmental Response, Compensation and Liability Act of 1976 and the Oil Pollution Act of 1990, there is no basis for estimating the magnitude of the savings; they will be excluded from this discussion. There are no known incentive-based programs for drinking water at present.

Effluent Charge Systems

Effluent charges are common in the United States and elsewhere. Dischargers to publicly owned treatment plants face charges that are based on volume or on volume multiplied by toxicity weights. In 28 states, the National Pollutant Discharge Elimination System (NPDES) permit fees for industrial dischargers to surface waters are based on volume or on volume and toxicity (Duhl 1993).

Very limited information is available concerning the impact of these fee systems on the volume or toxicity of industrial discharges. Simms (1977) found that pollutant-based charges provide an incentive for large industrial facilities to reduce effluents. Several studies cited earlier in this paper have modeled the potential savings from effluent charge systems, finding possible savings of 20% to 50% or more. Since charges for NPDES permits are based on volume and/or toxicity in approximately one-half of the states and sewerage charges are ubiquitous, the national magnitude of savings from current effluent charge systems could amount to 10% of point source discharge costs, or some $4.5 billion currently and $5.5 billion by the year 2000.

Reporting Requirement

As noted above, the 1986 Emergency Planning and Community Right-to-Know Act (SARA Title III) requires firms that manufacture or process in excess of 25,000 pounds of some 302 chemicals (currently) to report all quantities released to air, water, and public sewage facilities and deposited on land and injected underground. While not a water pollution control statute per se, SARA Title III requirements may influence the quantity of effluents released to the nation's waters. Between 1988 and 1993, reported surface-water discharges of toxic chemicals declined by 13%, from 312 million to

271 million pounds, for an average annual decline over this period of 2.6%. During the same period, reported discharges to public sewage facilities declined 45%, from 581 to 314 million pounds, for an average annual decline over this period of 9.2% (U.S. Environmental Protection Agency, 1995).

To estimate the potential reduction in annual compliance costs for controlling water discharges by the year 2000, we rely on these trends to project impacts at the rate of 2.6% per year for surface-water discharges and 9.2% for discharges to public sewage facilities. This results in cumulative reductions of about 17% and 49% by the year 2000, respectively, relative to 1993 levels. This would imply a reduction of about 200 million pounds of total toxic chemical releases by the year 2000 relative to a 1993 baseline, for an average annual decline of about 30 million pounds. Per-pound control cost savings are not likely to be as large as they were for air.[6] Consequently, projected savings from Title III reporting requirements relative to a CAC approach for water discharges are likely to be less than $10 million annually in the year 2000.

Trading Programs

Water effluent trading in the United States is currently confined to five separate programs. Three water effluent trading programs are designed to gain efficiencies in controlling nutrient discharges among point and nonpoint sources. Dillon and Cherry Creek reservoirs in Colorado and the Tar-Pamlico basin in North Carolina are examples. Wisconsin has a program that allows trading of effluent discharge credits between point sources. Finally, certain plants in the iron and steel industry are allowed to treat discharges from different points within the facility as if they originated at a single point, a so-called "water effluent bubble."

The only one of these programs that could be producing significant savings in compliance costs today is the iron and steel effluent bubble. However, as the provision is written, the bubble is believed to be applicable to only four plants in the Midwest, limiting potential savings to a few million dollars annually. The other programs have experienced very limited trading: Three trades have taken place at Dillon, one or two on the Fox River in Wisconsin, and none yet at

---

[6]The principal chemicals discharged to surface water are ammonium sulfate, phosphoric acid, methanol, and sulfuric acid. The three largest categories of chemicals discharged to public sewers are ammonium sulfate, methanol, and sulfuric acid. Control costs for all these substances except methanol are likely to be relatively modest; for example, acids can be neutralized with limestone at low cost.

Cerry Creek and in the Tar-Pamlico basin.[7] The U.S. Environmental Protection Agency strongly promotes the use of effluent trading to achieve water quality objectives and standards, including intra-plant trading, pretreatment trading, point/point source trading, point/nonpoint source trading, and nonpoint/nonpoint source trading. Trading is being explored, developed, or implemented in a number of watersheds throughout the country, in addition to those specifically mentioned here. The potential of trading, however, remains largely untapped.

In the aggregate, compliance cost savings from all existing water trading programs are currently small enough to have no discernible impact on the aggregate estimate for all incentives. In the late 1990s, annual savings on the order of $10 million appear possible from current effluent trading programs.

Summary of Cost Savings from Existing Incentives to Reduce Water Pollution

Effluent charges in the form of sewerage fees and fees for state NPDES permits are likely to be the most important source of savings from existing incentives for water pollution control, accounting for an estimated $4.5 billion in current savings and $5.5 billion in savings by the year 2000. Reporting requirements produce insignificant savings at present; but could yield cost savings of $10 million annually by the year 2000. Liability rules could produce major cost savings; however, there appears to be no means to estimate their potential impact. Effluent trading provisions may save $10 million or more annually by the year 2000, particularly if these mechanisms are more widely implemented.

Waste Management Control Programs

A number of incentive systems currently operate in the area of solid and hazardous waste. Over 2000 communities have instituted per-bag or per-can charge systems for household solid waste disposal. In addition, beginning with Oregon in 1972, approximately ten states now have deposit systems for beverage containers. Since savings from these deposit systems are likely to be small (see Porter (1978), for example), we do not estimate them below.

[7]Private conversation, U.S. EPA, Office of Water, 1995.

In the last three years, at least ten states have enacted deposit legislation concerning lead automotive batteries. A number of states are considering recycled content standards for newsprint as an incentive to stimulate newsprint recycling. We currently have no basis for estimating the magnitude of the cost savings from such programs and do not discuss them below.

SARA Title III reporting appears to be influencing the reported quantities of hazardous waste disposed of on land and injected underground. Private disposal charges for hazardous waste vary with quantity, providing an incentive to reduce wastes. Unfortunately, the impact of hazardous waste disposal fees has yet to be analyzed.

Marginal Cost Pricing for Household Waste

Cities that have adopted per-can or per-bag charge systems for household solid waste report significant decreases in the volume of waste collected. A representative figure is the 30% reduction observed in Seattle, when per-can fees were coupled with increased opportunities for households to recycle. The entire 30% does not represent a cost saving, however, since households experience some additional costs (primarily time) when they compost yard waste and recycle paper, aluminum, and glass. Cities obtain recycled commodities that have some economic value, offsetting some, but probably not all, of the increased cost to households. Thus, we estimate the net cost savings for the 2200 cities with per-can or per-bag systems to be in the range of 20%. Currently, we estimate that this may produce savings of approximately $100 million annually.

Reporting Requirements

In response to SARA Title III requirements, firms reported on-site land disposal of 513 million pounds of hazardous waste in 1988 and 289 million pounds in 1993, a decrease of nearly 44%, for an average annual decline of 8.7%. Reported underground injection of hazardous waste also fell from 1342 million pounds in 1988 to 576 million pounds in 1993, a decrease of 57%, for an average annual decline of 11.4%.

The magnitude of cost savings depends largely on the magnitude of present and future disposal costs. Land disposal of hazardous waste currently costs from about $50 to $200 per cu yd, or about 10 cents per pound, Underground injection is believed to be somewhat less expensive. It seems reasonable to assume that reductions in reported quantities achieved through requirements of SARA Title III cost on the order of one-half as much as would a CAC regulation that accomplished the same result. Thus, we estimate that, in 1993, reporting requirements saved industry about $6 million. Future disposal costs could rise as a result of the combination of a growing volume of trash and a declining number of landfills. In addition, new landfills may be much more expensive to operate as a result of regulatory requirements. Consequently, we estimate that cost savings associated with reporting requirements could rise to approximately $30 million annually by the year 2000.

Hazardous Waste Pricing Incentives

Disposers of hazardous waste (with the exception of small generators) must comply with requirements of the Resource Conservation and Recovery Act (RCRA). These requirements impose substantial costs on generators that vary directly with the volume of hazardous wastes produced. Hazardous waste pricing incentives are almost certainly impacting the volume of waste generated. Because generators are able to choose between paying for disposal or reducing their waste stream, pricing incentives will be more cost-effective than direct regulation and will result in cost savings for generators. Because this area has not been closely studied, there are no data to support a careful estimate. In the absence of any data, we assume that hazardous waste volumes are reduced by at least 25% through pricing incentives relative to a CAC system, indicating that cost savings could approach 25%, or approximately $1.8 billion annually in 1992, and $2.9 billion by the year 2000.

Summary of Cost Savings from Existing Incentives to Reduce Land Pollution

Existing solid waste incentive mechanisms are likely to produce savings of no more than $100 million per year currently; by the year 2000, these same incentives are unlikely to produce significantly greater cost savings unless they receive much wider use. In 1993, reporting requirements under SARA Title III saved industry about $6 million relative to CAC requirements. Cost savings associated with reporting requirements could rise to approximately $30 million annually by the year 2000.

Hazardous waste pricing incentives could be achieving significant reductions in current waste volumes, possibly as great as 50%. Unfortunately, no published studies of such impacts exist. In the absence of any data, it appears reasonable to assume that hazardous waste volumes are reduced by at least 25% through pricing incentives relative to a CAC system, indicating that cost savings could approach 25%, or approximately $1.8 billion in 1992, and $2.9 billion by the year 2000.

POTENTIAL SAVINGS FROM NEW INCENTIVE APPLICATIONS

Air Pollution Controls

This section uses two different methods for making a lower-bound estimate of the potential additional savings that would be possible through a wider use of economic incentive mechanisms for the control of air emissions, The first approach is to review studies that model incentive applications. Of the some 15 studies of this report summarized in Table 1.2.1, the *average* ratio of CAC costs to incentive-based costs is 5.2, indicating that, on average, an incentive approach would cost about 20% as much as a CAC approach. It would be far too optimistic to assume that savings of this magnitude could be achieved. Indeed, retrospective studies show that the actual savings from incentive applications have fallen well short of the theoretically predicted magnitudes as a result of unforeseen transaction costs and greater regulatory hurdles than were anticipated. Consequently, one might more realistically assume that, on average, incentive-based mechanisms could save 50% over CAC approaches for stationary source air pollution control. Mobile source programs might realistically achieve one-half that percentage saving, primarily from greater reliance on incentives to reduce pollution from existing vehicles rather than new vehicles.

The U.S. EPA (1990) estimates that air pollution control efforts will result in $43.5 billion in compliance expenditures by non-regulatory groups by the year 2000, consisting of $28.7 billion for stationary sources, $14.1 billion for mobile sources, and $0.7 billion for radiation. Using the assumption of average savings of 50% in compliance costs for stationary sources and 25% for mobile sources, widespread use of incentive-based programs might reduce compliance costs by $14.4 billion for stationary sources and $3.5 billion for mobile sources in the year 2000.

An alternative method that can be used to check whether these estimates are reasonable is to identify some major areas where incentive-based mechanisms could be applied and to calculate the potential savings from these applications. Out of necessity, this approach will neglect a host of relatively minor applications whose cumulative impact on costs could nonetheless be quite large. The discussion provides a rough indication of the magnitude of savings in compliance costs that may be achieved but does not purport to advocate any particular application of market incentives. As described in the previous section, existing (1992) savings from incentives to control air pollution are estimated to be in the range of about $1.7 billion (possibly somewhat higher because of failure to calculate a few items), rising to a range of $2.8 billion by the year 2000. Possible new applications described below could increase the annual savings by the year 2000 by an additional $12.6 billion. We rely on this more conservative estimate in our summary of cost savings.

Allow Trading Between Mobile and Stationary Sources

Evidence accumulated by the Congressional Budget Office (CBO), Freeman (1982), and others, suggests that, measured in terms of cost-effectiveness, mobile sources have been excessively controlled relative to stationary sources. White (1982) estimated that a reallocation of control costs could have saved the equivalent of $240 per vehicle, or about $2.5 billion annually.

Trading out of New Source Performance Standards (NSPS)

The 1977 Clean Air Act Amendments effectively require all new power plants to scrub their stack emissions of sulfur dioxide. The CBO estimates that, if utilities were allowed to meet NSPS requirements through the least-cost approach (such as trading with other sources and use of low-sulfur coal), the annual savings would be $4.2 billion per year.

Adopting RECLAIM in Other Areas

The northeast states, Houston, and other areas have considered creating trading regimes similar to RECLAIM. The SCAQMD projected that RECLAIM would produce savings in compliance costs for stationary sources on the order of 65% in the early years of the program. For this calculation, we reduce this to a 50% saving and apply that to EPA's 1990 estimate for the cost of controlling stationary source emissions of VOC and NO_X. Controls for these pollutants constitute about 30% of all air pollution control outlays for station-ary sources, or a projected $3 billion in 1992 and $4.5 billion in the year 2000. Savings of 50% of this would indicate potential savings in the range of $1.5 billion in 1992, rising to $2.2 billion in the year 2000, including $300 million estimated for RECLAIM in that year.

Fees on In-Use Emissions

Inspection and maintenance (I/M) programs suffer from incen-tives that encourage motorists and inspectors to falsify a vehicle's actual emissions, as well as from difficulties in correctly repairing problems that are identified. While EPA's new I/M requirements make cheating more difficult, an incentive system based on taxing a vehicle's actual in-use emissions would greatly reduce the opportun-ity to cheat. The potential savings might be estimated as follows: We assume that vehicle hydrocarbon emissions, currently some 6.5 million tons annually (Office of Technology Assessment 1989), could be reduced by 20% at a cost of $2,000 per ton rather than the $5,000 of a direct regulatory approach (the cost of basic I/M programs, estimated by EPA).[8] The savings associated with an emissions tax could approximate $4 billion annually.

WATER POLLUTION CONTROL

At least eight published studies listed in Table 1.2.1 show potential savings from using incentive mechanisms to control water pollution. Some of the studies consider hypothetical effluent fees, while others consider the potential impact of effluent credit trading systems. The ratio of CAC costs to those that would be incurred with an incentive approach ranged from a low of 1.12 to a high of 3.13.

For a variety of reasons, effluent trading systems are unlikely to be implemented widely. The most important constraint appears to be the limited number of bodies of water for which there are several dischargers of the same pollutant (U.S. EPA 1992, U.S. General Accounting Office 1992). Trading between municipal wastewater-treatment plants and nonpoint agricultural sources does appear to

---

[8]Independent analysts have estimated much poorer cost-effectiveness for basic I/M programs. Anderson and Lareau (1992) conclude that the cost-effectiveness of basic I/M ranges from $16,000 per ton to $35,000 per ton of hydrocarbons.

have the potential for relatively large cost savings, perhaps on the order of one-third to one-half of projected municipal wastewater expenditures of about $6 billion annually through the 1990s.

If there are to be further major savings in compliance costs from an incentive mechanism for water pollution, the most likely candidate mechanism is an effluent discharge fee. If one assumes average savings of one-third of normal CAC costs, somewhat less than the 40% to 50% average savings shown in theoretical studies, and applies this to *total* pollution control costs for all point source dischargers of $50.7 billion through the 1990s to the year 2000, one would estimate a potential saving in environmental costs of approximately $17.2 billion per year. (Approximately $5.5 billion of this total results from existing effluent charges, resulting in an additional $11.7 billion possible by 2000 from wide use of effluent discharge fees.)

Nonpoint source incentives could produce some savings if a significant nonpoint program is enacted into law. Under present law, nonpoint control costs will remain modest through the year 2000. Drinking water regulations might be amenable to incentive-based mechanisms, but the specifics remain unclear. Consequently, no savings are projected for this cost category.

SOLID AND HAZARDOUS WASTE CONTROL

Analysts have offered a number of incentive-based proposals for improving waste management policy.[9] Many of the suggestions involve more widespread application of incentives in current use: marginal cost pricing of household solid waste disposal, and container and battery deposits. Other suggestions are more novel: for example, virgin material content taxes, product tax/recycling subsidy systems, and recycling credits combined with recycled content standards.

For solid waste disposal, it appears reasonable to project that incentive mechanisms could achieve a 10% reduction in costs by the year 2000, a saving of $2.2 billion, based on assumed savings of 20% in about one-half of the nation. It appears likely that marginal cost pricing would not be cost-effective in some large cities and rural areas because of an inability to enforce sanctions against littering.

[9]See, for example, Wirth and Heinz (1991).

SUMMARY OF COST SAVINGS

The cost savings are summarized in Table 1.2.3. The table puts the savings into perspective relative to the total national costs of pollution control in the air, water, and land media. Nonregulatory costs are used as the basis for the estimates because there are unlikely to be any significant savings in terms of the expenses of government environmental regulatory agencies, which will still have to promulgate regulations governing the use of incentives and the enforcement of these regulations. Such savings for 1992 are estimated to have been about 7.6% ($8 billion) relative to compliance costs without incentives. Cost savings are projected to increase slightly to 7.7% ($11 billion) by the year 2000. When the projected savings in 2000 from existing incentives are added to the possible savings in 2000 from expanded use of incentives, total potential savings would be 26% (nearly $38 billion) of estimated compliance costs in the year 2000. Total potential savings from market incentives to control air pollution constitute the highest percentage of estimated compliance costs, at 35% of air pollution control costs.

Table 1.2.3. Summary of Estimated Cost Savings from use Economic Incentives (in Billions of Dollars)

Media	(1) Estimated environmental compliance costs[a] in 1992	(2) Savings due to existing incentives (% of col. 1)	(3) Estimated environmental compliance costs[a] in 2000	(4) Projected savings in 2000 from existing incentives (% of col. 3)	(5) Total potential savings in 2000 with existing and expanded incentives (% of col. 3)
Air	$28.9	5.9%	$43.5	6.4%	35.4%
Water	$45.1	10.0%	$57.4	9.6%	30.0%
Land	$32.6	5.8%	$45.6	6.6%	11.2%
Total	$106.6	7.6%	$146.5	7.7%	25.7%

[a]Environmental compliance costs exclude those costs of government environmental regulatory agencies that are predominantly related to promulgation and enforcement of regulation. These costs are not likely to be reduced through increased use of economic incentives. Estimates of compliance costs are in billions of 1986 dollars annualized at 7%.

Source: Estimated compliance cost estimates are from U.S. EPA (1990). Estimated savings estimates are explained in the text.

STIMULUS TO LONG-RUN INNOVATION

Even if cost savings from using market incentives are less than predicted as a result of regulatory, institutional, transaction, and/or legal constraints, the actual savings are still significant — especially when the stronger stimulus for innovation and technical change that the incentives provide is taken into account. In addition to offering cost savings, market-based approaches to environmental protection create greater financial incentives for the development of long-run innovations to control pollution than result from traditional regulatory approaches. When a regulatory approach with uniform requirements is used, utilities have little incentive to develop technologies to make emissions reductions below their tonnage limits. If firms reduce emissions below allowable levels with a command-and-control system, they receive no financial reward for doing so. In addition, regulations that contain technology-based standards can lock in the use of approved approaches, making it difficult for firms to get alternative approaches approved and paralyzing the development of more efficient new technologies. In contrast, market-based approaches provide financial rewards to firms that reduce releases below required levels, thereby building in a stimulus for the development of new technologies and innovations to achieve these results.

Illustration: Acid Rain Trading Program

The acid rain trading allowance program offers one example of the potential for innovation resulting from the use of market incentives. The acid rain trading program is expected to generate improved pollution control technology in the long run. Already, the cost associated with cleaning up a ton of SO_2 has started falling. Title IV under the 1990 Clean Air Act establishes a market-based approach to controlling acid rain. The system is designed to allow utilities to choose cost-effective pollution controls and thereby reduce the cost of meeting emissions limits. Under Title IV, EPA annually allocates a specific number of emissions "allowances" to utilities, each of which allows them to emit a ton of SO_2 during or after a specified year. At the end of the year, each utility must have one allowance for each ton of SO_2 emitted. Utilities have flexibility under Title IV to choose from a variety of cost-minimizing approaches to reduce emissions to the lower maximum SO_2 limits. The options available include switching to lower-sulfur coal, retiring an old plant, and installing pollution control devices. In addition, utilities can buy and sell SO_2 allowances. The system is designed to allow a utility to reduce compliance costs by purchasing additional allowances to satisfy requirements rather than further reducing emissions if its cost to reduce SO_2 emissions is higher than the market price of allowances. A utility with lower costs of emissions reductions can reduce emissions below the required level and sell surplus allowances to utilities with higher compliance costs.

The flexible range of various possible approaches to compliance with Title IV is resulting in lower prices for low-sulfur coal, scrubbers and allowances as vendors compete to meet the compliance needs of utilities. Prices of low-sulfur coal have fallen from an expected $40 per ton to under $25 per ton in 1995. Prices of scrubbers have also fallen by up to 50% since l990 (U.S. General Accounting Office 1994, p. 28). The reduced demand for scrubbers resulting as utilities choose alternative approaches of compliance has provided an incentive to manufacturers of scrubbers to innovate to reduce costs. They have developed larger absorbers, new anticorrosive materials, and methods of eliminating waste streams from scrubbers by converting them into marketable products (U.S. General Accounting Office 1994, pp. 25–29). Allowance prices reflect the falling costs of using low-sulfur coal and scrubbers resulting from such innovations. The falling allowance price reflects the declining marginal abatement costs associated with cleaning up a ton of SO_2. Although few trades between utilities have yet taken place, the ability of utilities to sell allowances generated by emissions control techniques that bring emissions well below targets provides companies with the incentive to further develop new and more effective emissions control technologies.

Illustration: Stratospheric Ozone Protection Program

To reach its goal of ozone protection, the stratospheric ozone protection program uses two market-based instruments: allocation and trading of chlorofluorocarbon (CFC) production allowances, and a tax on CFCs. As briefly described earlier, the EPA established a phaseout system for CFCs in 1991. To implement the system, the agency apportioned baseline allowances, provided a schedule for reducing the allowances, and allowed for trading of allowances among firms. A firm's production base declines in proportion to transfers in production allowances, plus an additional 1% of the amount traded (Congressional Research Service 1994, p. 71). Chlorofluorocarbon taxes were introduced in 1990 and were revised and extended in subsequent years. The taxes were designed to remove windfall profits that otherwise might accrue to producers as CFCs were phased out and to provide an incentive to develop substitutes in a timely fashion. Without the taxes, producers might respond to ever-tighter production quotas by simply raising prices. In addition, the incentives to develop substitutes might be weak since creating and producing substitutes could lower industry profitability. Thus, the production of CFCs is impacted by a tax, which increases over time, and a production quota, which decreases over time.

Reliance on market mechanisms has provided maximum flexibility to CFC manufacturers, allowing CFC users to devote resources to the development of innovative transition technologies and alternative substances, with the result that CFC production has declined much more rapidly than anticipated. In 1987, the optimistic outlook was for a decline in CFC production to 50% of 1986 levels by 1999. By the end of 1991, CFC production was already down to 60% of 1986 levels; by the end of 1992, it was down to less than 50% of 1986 levels (Congressional Research Service, 1994, p. 73). Although there has been little activity under the CFC allowance trading system, one factor contributing to this phenomenon is most probably the rapid growth of substitutes, which has contributed to production of CFCs in excess of market needs.

Responses to the market-based programs to control CFC production have varied across industrial sectors. The printed circuit board (PCB) manufacturers in the electronics industry, including Motorola, AT&T, and Northern Telecom, used a cooperative approach to respond to the phaseout. The EPA assisted the industry in forming the Industry Cooperative for Ozone Layer Protection (ICOLP), capitalizing on the interest of these companies in collaboration. The ICOLP provided companies with a forum for sharing information on innovative approaches to reduce and eliminate dependence on CFCs for cleaning PCBs. The consortium investigated first the aqueous cleaning method and then the "no-clean" process. This phased approach to technology innovation would have been much less likely under a more traditional regulatory scheme. As a result of the success of the new technology, ICOLP mounted a widespread effort to promote and disseminate to other companies information on no-clean processes. Many companies eventually set corporate phaseout dates for CFC use that were even more aggressive than the phaseout dates set by EPA.

The foam-blowing industry has a parallel effort in industry cooperation. All major U.S. foam-blowing companies that manufacture foam-insulated home appliances are now members of a subsidiary of the American Home Appliance Manufacturers established by the companies to research and share technological innovations for reducing dependence on ozone-depleting chemicals. The rising cost of CFCs, resulting from the decreases in production volume and the CFC tax, has made most substitute foam-blowing agents competitive with CFCs. The tax has expedited the adoption of substitutes in many electronics and other cleaning operations (Congressional Research Service 1994, p. 74).

Summary

The experiences of the acid rain allowance trading program and the stratospheric ozone protection program begin to illustrate the potential for technological innovation from the use of incentive-based pollution control programs. The stimulus for innovation and technical change offers an important supplement to the opportunity for cost savings resulting from the use of market-based programs.

CONCLUSION

Although the figures presented here fall far short of providing a definitive dollar estimate of potential cost savings from the broad use of market incentives in the United States, they illustrate the magnitude of savings that might be achieved and are meant to foster further study and debate over the expanded use of market-based approaches to control pollution. With the prospect of enhanced technological innovation accompanying their use, the advantages are compounded. The potential for cost savings and the opportunities for increased innovation resulting from broad use of market-based approaches to pollution control suggest that this approach remains favorable to regulation and subsidization, where practicable, as J. H. Dales concluded in 1968. Through these advantages, market-based approaches offer a potentially rewarding economic opportunity as well as continued progress in protecting the environment. While the implementation of new incentive programs requires flexibility and the overcoming of uncertainty and obstacles to enforcement, the potential rewards make the effort worthwhile.

REFERENCES

Anderson, Robert, C., and Thomas J. Lareau. 1992. *The Cost Effectiveness of Vehicle Inspection and Maintenance Programs.* Research Study 067. Washington, DC: American Petroleum Institute.

Anderson, Robert C., Lisa A. Hofmann, and Michael Rusin. 1990. *The Use of Economic Incentive Mechanisms in Environmental Management.* Research Paper 051. Washington, DC: American Petroleum Institute.

Atkinson, Scott E. 1983a. "Marketable Pollution Permits and Acid Rain Externalities." *Canadian Journal of Economics 816:708–*22.

Atkinson, Scott E. 1983b. "Nonoptimal Solutions Using Transferable Discharge Permits: The Implications of Acid Rain Deposition." In *Buying a Better Environment*, edited by E. F. Joeres and M. H. David. Madison, WI: University of Wisconsin Press.

Atkinson, Scott, E., and Donald H. Lewis. 1974. "A Cost-Effectiveness Analysis of Alternative Air Quality Control Strategies." *Journal of Environmental Economics and Management* 1:237–50.

Atkinson, Scott, and Thomas Tietenberg. 1991. "Market Failure in Incentive-Based Regulation: The Case of Emissions Trading." *Journal of Environmental Economics and Management* 21:17–32.

Burtraw, Dallas. 1995. "Cost Savings Sans Allowance Trades? Evaluating the SO_2 Emission Trading Program To Date." Discussion Paper 95-30. Washington, DC: Resources for the Future. September.

Carlin, Alan. 1992. *The United States Experience with Economic Incentives for Environmental Pollution Control.* Report EPA-R-92-001. Washington, DC: U.S. Environmental Protection Agency.

Clinton, Bill, and Al Gore. 1995. "Reinventing Environmental Regulation." Unprocessed, White House. Washington, DC: (March 16).

Congressional Budget Office. 1982. *The Clean Air Act, the Electric Utilities, and the Coal Market.* Washington, DC: U.S. Government Printing Office.

Congressional Research Service. 1994. *CRS Report for Congress: Market-Based Environmental Management: Issues in Implementation.* The Library of Congress, Environment and Natural Resources Policy Division, 94–213 ENR. March 7.

Dales, J. H. 1968. *Pollution, Property and Prices.* Toronto, ON: University of Toronto Press.

David, Martin, Wayland Eheart, Erhard Joeres, and Elizabeth David. 1977. "Marketable Effluent Permits for the Control of Phosphorous Effluent in Lake Michigan." Social Systems Research Institute Working Paper, University of Wisconsin, Madison, WI. December.

Dudek, Daniel, and John Palmisano. 1988. "Emissions Trading: Why Is This Thoroughbred Hobbled?" *Columbia Journal of Environmental Law* 13:217–256.

Duhl, Joshua. 1993. *Effluent Fees: Present Practice and Future Potential.* Discussion Paper 075. Washington, DC: American Petroleum Institute.

Eheart, Wayland, E. Downey Brill Jr., and Randolph M. Lyon. 1983. "Transferable Discharge Permits for Control of BOD: An Overview." In *Buying a Better Environment: Cost-Effective Regulation through Permit Trading*, edited by Erhard F. Joeres and Martin H. David. Madison, WI: University of Wisconsin Press.

Freeman, A. Myrick III. 1982. *Air and Water Pollution Control: A Benefit-Cost Assessment,* New York: Wiley.

Hahn, Robert. 1989. "Economic Prescriptions for Environmental Problems: How the Patient Followed the Doctor's Orders." *Economic Perspectives* 3(2):95–114.

Hahn, Robert. 1995. "An Economic Analysis of Scrappage." *Rand Journal of Economics* 26(2):222–242. Summer.

Hahn, Robert and Gordon Hester. 1989. "Where Did All the Markets Go? An Analysis of EPA's Emissions Trading Program." *Yale Journal on Regulation* 6(1):109–53.

Hahn, Robert, and Roger Noll. 1982. "Designing a Market for Tradable Emissions Permits." In *Reform of Environmental Regulation,* edited by Wesley Magat. Cambridge, MA: Ballinger, pp. 132–133.

Harrison, David Jr. 1983. "Case Study 1: The Regulation of Aircraft Noise." In *Incentives for Environmental Protection,* edited by Thomas C. Schelling. Cambridge, MA: MIT Press.

ICF Incorporated. 1991. *Regulatory Impact Analysis of the Proposed Acid Rain Implementation Regulations.* Report EPA-400-1-91-049, prepared for the Office of Atmospheric and Indoor Air Programs, Acid Rain Division, U.S. EPA. September 16.

ICF Resources International. 1989. *Economic, Environmental, and Coal Market Impacts of SO_2 Emissions Trading Under Alternative Acid Rain Control Proposals.* Report prepared for the Regulatory Innovations Staff, Office of Policy, Planning and Evaluations. Washington, DC: U.S. EPA.

Johnson, Edwin L. 1967. "A Study in the Economics of Water Quality Management." *Water Resources Research* vol. 3, issue 2:291–263.

Kohn, Robert E. 1978. *A Linear Programming Model for Air Pollution Control.* Cambridge, MA: MIT Press.

Krupnik, Alan. 1986. "Costs of Alternative Policies for the Control of Nitrogen Dioxide in Baltimore." *Journal of Environmental Economics and Management* 13:189–197.

Liroff, Richard. 1986. *Reforming Air Pollution Regulation: The Toil and Trouble of EPA's Bubble.* Washington, DC: The Conservation Foundation.

Maloney, Michael T., and Bruce Yandle. 1984. "Estimation of the Cost of Air Pollution Control Regulation." *Journal of Environmental Economics and Management,* vol. 11, issue 3:244–263. September.

McGartland, Albert M. 1984. "Marketable Permit Systems for Air Pollution Control: An Empirical Study." Ph.D. Dissertation, University of Maryland, College Park, MD.

Nichols, Albert L. 1983. "The Regulation of Airborne Benzene." In *Incentives for Environmental Protection*, edited by Thomas C. Schelling. Cambridge, MA: MIT Press.

Oates, Wallace E., Paul R. Portney, and Albert M. McGartland. 1989. "The Net Benefits of Incentive-Based Regulation: A Case Study of Environmental Standard Setting." *American Economic Review* 74:1223–1242.

Office of Technology Assessment. 1992. *Retiring Old Cars: Programs to Save Gasoline and Reduce Emissions,* OTA-E-536. Washington, DC.

Office of Technology Assessment. 1989. *Catching Our Breath: Next Steps for Reducing Urban Ozone,* OTA-0-412. Washington, DC.

O'Neil, William B. 1980. "Pollution Permits and Markets for Water Quality." Ph.D. Dissertation, University of Wisconsin-Madison, WI.

Opaluch, James J., and Richard M. Kashmanian. 1985. "Assessing the Viability of Marketable Permit Systems: An Application in Hazardous Waste Management." *Land Economics* 61:263–271.

Palmer, Adele R., William E. Mooz, Timothy H. Quinn, and Kathleen Wolf. 1980. *Economic Implications of Regulating Chlorofluorocarbon Emissions from Nonaerosol Applications.* Report R-2524-EPA. Santa Monica, CA: The RAND Corporation. June.

Porter, Richard C. 1978. "A Social Benefit-Cost Analysis of Mandatory Deposits on Beverage Containers." *Journal of Environmental Economics and Management* 5:351–375.

Roach, Fred, Charles Kolstad, Allen V. Kneese, Richard Tobin, and Michael Williams. 1981. "Alternative Air Quality Policy Options in the Four Corners Region." *Southwest Review* 1(2):44–45.

Schroeer, William L. 1992. "A Cost-Effective Accelerated Scrappage Program for Urban Automobiles," presented to Transportation Research Board (unpublished). January 13.

Seskin, Eugene P., Robert J. Anderson Jr., and R. O. Reid. 1983. "An Empirical Analysis of Economic Strategies for Controlling Air Pollution." *Journal of Environmental Economics and Management* 10:112–124.

Shapiro, Michael, and Ellen Warhit. 1983. "Marketable Permits: The Case of Chlorofluorocarbons." *Natural Resources Journal* 23:577–591.

Simms, William A. 1977. *Economics of Sewer Effluent Charges.* Ph.D. Thesis, Department of Political Economy, University of Toronto, ON.

South Coast Air Quality Management District (SCAQMD). 1992. *Regional Clean Air Incentives Market.* Diamond Bar. CA. SCAQMD. Spring.

Spofford, Walter O. Jr. 1984. "Efficiency Properties of Alternative Source Control Policies for Meeting Ambient Air Quality Standards: An Empirical Application to the Lower Delaware Valley." Unpublished, Resources for the Future Discussion Paper D-118. February.

Tietenberg, T. H. 1985. *Emissions Trading: An Exercise in Reforming Pollution Policy.* Washington, DC: Resources for the Future.

U.S. Environmental Protection Agency. 1985. "Regulation of Fuels and Fuel Additives; Banking of Lead Rights." vol. 50 *Federal Register* 13116–28. April 2.

U.S. Environmental Protection Agency. 1990. *Environmental Investments: The Cost of a Clean Environment, Report of the Administrator of the Environmental Protection Agency to the Congress of the United States.* Report EPA-230-11-90-083. Washington, DC. November.

U.S. Environmental Protection Agency. 1991a. "Air Programs; Credit Program for California Pilot Test Program." *Federal Register,* vol. 56, No. 186. 48618. September 25.

U.S. Environmental Protection Agency. 1991b. "Protection of Stratospheric Ozone," *Federal Register,* vol. 56, No. 189. 49548–580. September 30.

U.S. Environmental Protection Agency. 1991c. "Protection of Stratospheric Ozone," *Federal Register,* vol. 56, No. 250. 67368. December 30.

U.S. Environmental Protection Agency. 1992. *The Benefits and Feasibility of Effluent Trading Between Point Sources: An Analysis in Support of Clean Water Act Reauthorization.* Report EPA-68-WI-009. Washington, DC, May.

U.S. Environmental Protection Agency, Office of Pollution Prevention and Toxics. 1995. *1993 Toxics Release Inventory*, Report No. EPA-745-R-95-010. Washington, DC. March.

U.S. General Accounting Office. 1994. "Air Pollution: Allowance Trading Offers an Opportunity to Reduce Emissions at Less Cost." Report to the Chairman, Environment, Energy, and Natural Resources Subcommittee, Committee on Government Operations, House of Representatives. GAO/RCED-95-30. December.

White, Lawrence J. 1982. *Regulation of Air Pollutant Emissions from Motor Vehicles*, Washington, DC: American Enterprise Institute.

Wirth, Senator Timothy, and Senator John Heinz. 1991. *Project 88— Round II: Incentives for Action: Designing Market-Based Environmental Strategies*. Alliance to Save Energy. Washington, DC. May.

Part

2

❑

REGULATORY INNOVATIONS

ON TRIAL:

ENVIRONMENTAL MARKETS

❑

2.1

❏

INTRODUCTION TO PART 2:
FROM JOURNAL ARTICLES TO ACTUAL
MARKETS: THE PATH TAKEN

The Editors

❏ Tradable rights to pollute and the markets in which they may be exchanged occupy an important place in this volume. Specific types of tradable permits and particular markets undergoing development are the subject matter of the studies in this part. A general background to these detailed treatments is indispensable, in the editors' view, to understanding the current widespread interest in making use of this regulatory instrument. Since no one study undertakes this task, the editors endeavor to supply some relevant history and a few major definitions in this introduction.

The idea of tradable emission permits, or tradable pollution concentration reduction credits, lived a quiet life in academic journals for much of the post–World War II period.[1] There it received the attention of researchers who found, given simplifying assumptions, that the idea could lead to efficient control of environmental problems. It was a quiet life compared to the surrounding political debate on environmental policies and policy instruments that began in earnest about 1970. That debate, driven, in our view, by a growing concern for reducing the impacts of pollution as quickly and directly as seemed possible and by a lingering Depression-fueled distrust of

---

[1]Early research that laid a rigorous foundation for the analysis of cost-effectiveness of tradable permits was carried out by W. David Montgomery (1972). T. H. Tietenberg (1985) supplies a survey of theoretical work.

49

markets, gave rise to what came to be called *command-and-control (CAC) regulation*. A good example is found in the measures required by the Clean Air Act of 1970.

After 1970, the continuing but slow improvement in environmental quality that was more and more evident required expenditures that grew as a percentage of gross domestic product (GDP). Despite these expenditures and noticeable environmental improvement, many old problems, such as photochemical smog and acid rain, remained intractable, and several new problems, such as global climate change, loomed on the horizon. The initial dominant form of government regulation—the specification of uniform emission or effluent standards or particular technologies for all sources of pollution—was challenged in a growing number of studies as being cost-ineffective and as stifling innovation. At the same time, distrust of markets was greatly dissipated by the performance of the economy. The issue of the appropriate selection among policy instruments, more so than the level of the goals or targets, moved higher on the political agenda. These developments came together in the Clean Air Act Amendments of 1990. This legislation, for the first time, mandated an environmental market for the control of one pollutant and allowed environmental markets as an option for the control of others.

EARLY EXPERIMENTS IN TRADING

ERC Programs

These trends toward greater confidence in markets and the rising share of CAC expenditures in GDP did not pass unnoticed prior to 1990 in either the regulated or regulating communities or among many environmental and public interest groups. The U.S. EPA had experimented with the introduction of alternative, more decentralized incentive schemes, which resulted in a limited market restricted by emission trading rules. These are the offset, bubble, or netting programs which provide opportunities for emitters to earn emission reduction credits (ERCs), the tradable commodity. Emitters can use any reduction control practice or technique they want, provided that pollution is reduced below the current regulated amount. By allowing emitters to make their own selection of control practices or equipment in order to earn credits, some flexibility was introduced into the regulatory process. These programs, undoubtedly, have resulted in savings, as estimated in our first study, but there have been a number of constraints on their performance.

In order to participate in these programs, emitters must apply to the EPA for prior certification of credits and must demonstrate that emissions are truly reduced below the authorized level. Such credits must be shown to be surplus, enforceable, permanent, and quantifiable. Transactions can be affected (discounted) by the location of buyer and seller if the pollutant is considered to be nonuniformly mixed in the atmosphere—the "hot spot" problem. Emission reduction credits do not provide an effective means of reducing pollution but act to redistribute it, when trades occur, in a more cost-effective manner.

Balanced against these limitations, emission reduction credits have the advantage of fitting into, or on top of, traditional control measures. By reducing pollution below the regulated level, the emitter can sell credits to another emitter, who can now exceed the stipulated level. This is how the programs impart some flexibility to regulation. With this flexibility, cost-minimizing emitters choose the cheapest ways to reduce pollution and earn credits to be sold to emitters who cannot reduce cheaply. This is how the programs realized savings in the social costs of control of a given aggregate amount of pollution.

Considering both limitations and advantages, the conclusion must be reached that emission reduction credits have fallen short of attaining the full measure of savings possible. Although good data are hard to come by, the limitations on transactions, by all accounts, have reduced the number of possible trades. Of those made, many have been carried out by large emitters frequently "trading" within their own facilities or emission sources (Tietenberg 1985). The programs remain a distant relative to a well-defined market that has a freely exchangeable commodity whose performance can lead, in theory, to a least-cost solution of pollution control.

Lead Trading

Lead can be harmful and toxic to humans. Phasing out lead in gasoline presented an opportunity to design an incentive plan and to save on compliance costs since refineries had substantial differences in their costs of reducing lead, an important requisite for achieving cost-effectiveness by market-based approaches. Refineries doing better than the lead reduction standard during a quarter of a year would earn credits that could be sold to other refineries or banked. Given the tight deadlines established in the phaseout program, which put pressure on refineries to take rapid and sometimes expensive action, these tradable credits saved control expenditures, avoided litigation costs, and achieved the stated goal by 1987. Although the refinery industry is exceptional, with a few large facilities, the lead program is widely regarded as a successful use of incentives.

Ozone Depletion—The Montreal Protocol

Another development relevant to our purposes occurred as a result of international agreement—the Montreal Protocol and its subsequent amendments. The U.S. designed a tradable permit system as one component of its program to reduce and eventually eliminate certain stratospheric ozone-depleting substances such as choroflourocarbons. Tradable allowances were issued to both producers and major consumers based on 1986 use, and these periodic issuances of allowances were decreased over time, which is an important feature of a cap-and-trade market. The decline in allowances was scheduled to be steep, leading to a complete phaseout by the year 2000 if the current timetable holds. Because such a rate of decline could lead to large windfall profits due to an ever-declining stock of these substances so important in cooling and refrigeration, a special tax on profits of producers was imposed. Tradable credits are only a small part of the comprehensive elimination program, but their use has indicated growing interest and support for market-based approaches.

A FEW KEY DEFINITIONS

Types of Permits

Tradable emission permits or allowances to pollute are like private property in general, bundles of rights that confer on the owner certain privileges and responsibilities. The rights contained within an emission permit or allowance depend on the design of each particular trading program. They are not necessarily equal in their property rights. Most tradable commodities are denominated in the units of pollution—for instance, so many tons of sulfur dioxide or pounds of volatile organic compounds. They must be turned over to the regulating agency when that unit of pollution has been emitted.

A comprehensive bundle of rights confers on the owner the right to sell to or buy from any trader, including emitters, brokers, market makers, and the like, as in the case of SO_2. Less comprehensive bundles limit trading to emitters or subsets of emitters, as in the Illinois Environmental Protection Agency plan for control of urban ozone. Another element in the design of an allowance concerns dating: The tradable commodity can be dated for use during a limited period, banked for a limited period, or banked for as long as desired (SO_2). Tradable rights can be issued periodically or as a vector of permits usable at particular future dates or can be issued all at once (cumulative or evaporative permits) if a finite amount of pollution and no more is the goal. As will be seen in the coming discussion, the banking right has been resisted at times in fear of building up an accumulation of permits that could result in an emission "spike." The opposite problem is a price "spike," which could occur if banking is not allowed and emitters are forced to scramble for limited permits at the close of the permit period. Banking, it is further argued, can serve to clean-up the environment more quickly, given that a banked unit is not an emitted unit.

A controversy has developed around the proviso in most enabling legislation that limits the private property character of tradable permits so that the government can, without liability, alter the number of permits to be allocated or discount outstanding permits when new knowledge of the harms of pollution becomes available. That is, the quantity of permits can be altered without penalty to the regulating agency but with clear consequences to the permit holder, who stands to lose some fraction of a valuable asset. The threat of frequent changes in the quantity of permits would undoubtedly add a new uncertainty cost to the market. However, this proviso is similar to other limitations put on the use of private property that can also be put into effect at any time; for example, private property values may be affected by a change in zoning.

Types of Environmental Markets

We have already discussed the emission reduction credit programs, which create a limited market. Among the more comprehensive types of markets, two deserve special mention: the open market and the cap-and-trade or budget market. Cap-and-trade systems offer a way to establish an aggregate cap over the pollutant and to reduce that cap over time, thus giving the government control over the volume of pollution and replacing or making unnecessary additional CAC regulation. Allocations of tradable permits to individual sources are made under the aggregate cap. Thus, a comprehensive market can be established over all sources in which incentives can act to the fullest extent to furnish flexibility, to achieve cost-effective reductions, and to stimulate innovations. For successful implementation, these systems require information about the pollutants, the emission sources, and the distribution among sources of control techniques and costs. While this information may sound extensive, it is likely to be more limited in nature and extent than that dictated by CAC measures, which impose on the regulator the need for as much technical detail as the emitter possesses.

However, cap-and-trade systems do require detailed agreements between regulated and regulating communities that can complicate the application process where numerous business interests and local, state, and regional authorities are involved. Such markets require agreement on, and information about, the universe of emission sources to be covered and the baseline emissions levels from which reductions are to be achieved. They require agreement on, and information about, the aggregate cap and the individual allocations under that cap. They require agreement on the rate of decline of the cap, typically a matter of intense interest to each and every source. And they require agreement on standardized and effective monitoring and measuring techniques for determining and enforcing source emissions. Only recently has the technology for monitoring and measuring pollution in many industries come into being (or been developed under stimulation by regulation).

The U.S. EPA has proposed, as an alternative to the cap-and-trade system, an open market that would require fewer and less intricate agreements. The open market could provide a transitional form to the cap-and-trade model. It builds on the emission reduction credit programs previously described but simplifies and streamlines the regulatory procedures. For example, advanced approval of a transaction need not be secured by an emitter. The buyer of an open-market credit, however, must assure the regulating agency that the credit has been properly generated. The open market does not require a cap or rate of decline and thus does not give the regulating agency a means to affect the aggregate volume of pollution through this market approach. It also provides no assurance that the full measure of cost-effectiveness will be achieved.

A PREVIEW OF THE STUDIES

As a good example of how many of these issues just discussed arise in the actual design of a cap-and-trade market at the highly visible stage of its initiation, we have selected the study by Kanerva and Kosobud that describes a partial cap-and-trade market planned to control emissions of volatile organic materials (VOMs) or hydrocarbons in the severe nonattainment area of northeastern Illinois. On the grounds that the location of the ozone plume is shifting and is hard to predict, the design treats the ozone concentrations as uniformly mixed, thus simplifying the market and unifying the tradable allowances. The hot spot problem—in which a unit of emissions from one location has a greater impact on ozone formation in the nonattainment area than emissions from another—is to be monitored but not incorporated into the market design.

The market is partial in that only stationary sources of VOMs are covered. It is envisioned that the partial market could be extended by provisions enabling stationary sources to receive tradable allowances by reducing emissions from other sectors such as the mobile on-road sources. Credits could be earned, for example, from programs to scrap high-emitting cars or to convert buses to natural gas. Credits could also be earned from reducing emissions from the many small-area or off-road source—retiring two-stroke lawnmowers, for example.

Planning for this innovative regulatory measure in Illinois has stimulated another innovation in the way agreements and negotiations are reached on new environmental regulation. The institutional procedure in Illinois for introduction of a new environmental regulation has been, typically, for the Illinois EPA to formulate the proposal and submit it to a separate entity, the Illinois Pollution Control Board, at which time public hearings and serious negotiations between regulated and regulating communities get under way. As previously mentioned, cap-and-trade markets require a long list of negotiations and agreements. To speed up the process and raise the chances of successful implementation, the Illinois EPA began the negotiations and securing of agreements before submitting the proposal to the Pollution Control Board. By opening up these discussions to all concerned groups and by securing agreements on the rule-making language, the Illinois EPA hopes to avoid endless disputes.

Another feature of Kanerva and Kosobud's study that makes it of special interest is its description of the rapid translation of new scientific knowledge into the design of the market. The first cap-and-trade market was designed to control nitrogen oxide emissions. However, detailed modeling of the physical, chemical, and meteorological conditions in the region, as described in the study by Gerritson, found, in simulations of the model, that reducing nitrogen oxide (NO_x) emissions alone could increase ozone formation in parts of the nonattainment region. In the simulations, the reduction in the ozone cleansing effect of NO_x more than offset its ozone precursor effect. Moreover, it was found that the concentrations of precursors and ozone coming into the region were so high that unrealistic local reductions would be necessary to reach attainment. Attention was directed to the importance of new policies covering a larger area, perhaps a good part of the nation, to control these pollutants.

These discoveries led to a revised cap-and-trade market not for NO_x but for VOM control and to a state initiative to push for control outside the nonattainment area, both within and outside the state, of NO_x emissions, perhaps by use of tradable permits. A successful program in northeastern Illinois may well become a model for effective integration of scientific findings, political negotiations, and economic benefits of a market-based approach.

The Congressional stipulation of a cap-and-trade market for the control of sulfur dioxide, a precursor of acid deposition, was to generate a stream of regulatory rulemaking, controversy, research, and interest in further applications of this model that continues to the present day. The interesting story of the historic and hotly contested passage through the U.S. Congress of Title IV of the CAAA '90 is becoming clearer as William Rosenberg, assistant administrator for air and radiation of the U.S. EPA at that time, takes us behind the scenes for an account of the bargaining and negotiation that took place.

We then proceed to a study by the economist Ken Rose, who guides us through the mechanics of the SO_2 allowance market and the complications and restrictions connected with the electric utilities, which are its major players. This industry is historically and at present heavily regulated with respect to electricity rates and the rate base. Rose examines how this regulation may be affecting the operation of the cap-and-trade market.

The initial transitional Phase I, 1995–1999, of the market applies a cap to 72 of the heaviest SO_2 utility emitters east of the Mississippi River. Phase II beginning in the year 2000, applies a tighter cap (roughly half of the 1980s baseline) to 269 utilities nation wide. NO_x emissions, also a precursor of acid deposition, are currently controlled by more traditional regulation but, at the date of this writing, an association of state governments in the form of the Ozone Transport

Assessment Group are formulating plans for regional controls including a tradable NO_x emissions cap-and-trade market.

As we have noted earlier, the performance of the SO_2 market will be closely watched as a model for use elsewhere. Early data on prices and quantities of annually issued emission allowances, which can be bought, sold, or banked, are available from the auctions that have been held once a year since 1993. Managed for the U.S. EPA by the Chicago Board of Trade, these auctions have already made available valuable data for the years 1993 through 1996. Other private transactions can occur at any time outside the auction, but prices and quantities are much harder to discover. The auction has two parts: It makes available tradable permits, about 3% of the total, that are officially held back from emitting sources (to whom net revenues are returned), and it also makes available any allowances that are privately offered. The auction is a discriminant type of market, as required by Congress, that matches high bidder price to low asking price.

Several striking features of these early data have emerged. Few privately offered allowances have been sold. The prices of officially offered allowances, all of which have been sold, have been far below earlier expectations and, in the 1996 auction, the average price (weighted by quantity) was under $70 per ton. Furthermore, the prices of auctioned allowances dated for use six or seven years hence have also been low despite the reductions in store for Phase II. In a well-functioning market, the price of an allowance would equal, in equilibrium, the marginal control costs (properly discounted for allowances of future date) of reducing emissions. Few observers would have expected costs to be so low or to be lowered so quickly.

It is likely that the marginal control cost curve for SO_2 emission reductions has been shifted downward in response to the incentives of the market, but it is difficult to believe that this alone is responsible for the revealed low prices of the traded allowances. There are several puzzling factors that could be additional contributors to this result. Only a few utilities have engaged in the annual auction market, and anecdotal evidence suggests that private trades outside the auction have not been numerous. Are the utilities able to realize and take advantage of all the possible gains from trading allowances when so few transactions have been made? Sulfur dioxide emissions, in general, have been below the annual allocation of allowances, suggesting that many utilities are banking allowances and adding to sizable portfolios. It is conceivable that some utilities are reducing emissions using control measures whose costs are above the revealed allowance price in the auction.

As an economist studying utility behavior in general, Rose is in a good position to raise and discuss these issues. The slow pace of state and federal regulatory decisions on the treatment of allowances, the lower than expected prices for control inputs such as low-sulfur coal and SO_2 scrubbers, and the biases in the type of auction market specified by the U.S. Congress have added to the hesitancy of utilities to trade, and all these factors appear to have played a part in explaining the lower than expected prices. An ambitious future research program is in order to fully understand this market.[2]

The study that follows by Dudek, Goffman, and Wade contributes a survey of environmental market developments across the country and examines in depth the comparative advantages of the cap-and-trade model that may lead to its more frequent use in the future. Their survey reveals the potential range of applications, not all of which could be covered in this volume. Two of the authors are environmentalists who played an important role in the legislative process leading up to the adoption of the SO_2 market, and they continue to be active in providing advice on the appropriate utilization of this regulatory innovation. They are proponents of the use of cap-and-trade rather than open markets, and they present a thorough appraisal of the comparative merits of each.

George Tolley, in his study, prepares a list of possible transaction costs that could reduce the cost-effectiveness of tradable emission markets. On this basis, he suggests that emission taxes deserve a more respectful consideration as an alternative. While the editors retain their high regard for markets, Tolley's contribution influenced our decision to include in Part III a paper by Robert Repetto on green taxes in order to present a full range of control options to the reader.

It had been planned to include a study of the cap-and-trade markets under way and planned in the Los Angeles region managed by the South Coast Air Quality Management District (SCAQMD). Under the acronym of RECLAIM (Regional Clean Air Incentives Market), cap-and-trade markets for local sulfur dioxide, nitrogen oxides, and VOC emissions were initially drawn up by SCAQMD. The first two are under way; the latter is being revised. Dudek, Goffman, and Wade include an overview of the RECLAIM program in their survey. There were several reasons that worked against a more intensive Workshop inquiry into these markets.

[2]Many studies are under way, among them a recently published work by Burtraw (1996) that weighs the various reasons for low allowance prices and a study by Joskow and Schmalensee (1996) that carries out a detailed analysis of the factors behind the initial allocations of allowances to individual utilities.

In many respects, the NO_x market has been of greatest interest, having exhibited the largest number of transactions and having exhibited price paths for dated tradable permits that rise over time as the cap is racheted down (SCAQMD 1996a). Each emission source is allocated a vector of dated credits for the current year and for specific years through 2010. Credits can be used only for the year indicated but can be traded in advance. It is these credits with future dates that have shown the expected price increase because of the declining aggregate cap and individual allocations. While several studies of these transactions are under way and should provide a brighter illumination on the performance of this market (Prager et al. 1996), none were ready for publication at the time this book went to press.

The RECLAIM cap-and-trade market for local SO_2 emissions differs in many respects from the one established for the national scene by Title IV of the CAAA of 1990. RECLAIM tradable credits for SO_2 are allocated mainly to oil refineries in Southern California, which largely burn natural gas, not coal. Credits for use only in future years have sold for almost $1,000 per ton, reflecting the special control costs for refineries. Transactions have been limited in number.

The RECLAIM cap-and-trade market for VOCs has apparently been dropped because of disagreements between the regulating and regulated communities on benchmark emissions, the cap, and its rate of decline. An "open" type of market is on the drawing boards for consideration, but final agreements and approvals have not yet been secured.

The coexistence of "old-fashioned" emission reduction credits and newer RECLAIM tradable units has led the SCAQMD to propose a universal trading credit (UTC) that would create an umbrella market for control of each pollutant—namely, VOCs and NO_x, but possibly also carbon monoxide, and particulate matter (SCAQMD 1996b). Various types of credits could be exchanged for a UTC. In addition, an Air Quality Investment Program has been proposed that would enable emitters to make payments, in lieu of reductions, into a fund that would be used to secure reductions in emissions elsewhere. These proposals remain on the drawing board. Together with the maturing RECLAIM cap-and-trade markets for SO_x and NO_x, they deserve close examination in the near future.

Part 2 closes with a presentation by the editors of the results of a deliberative opinion poll taken of Workshop participants. The poll is not a random sample since the Workshop participants were not chosen to reflect the population at large, and its results cannot be projected to larger populations. The idea of the poll is that information combined with deliberation encourages reasoned judgments about environmental markets. Participant views might give an indication of what the general public would come to believe if they were so informed and had a similar opportunity.[3]

After the experience of Workshop meetings, the participants were positive in their rating of environmental markets and gave markets for control of acid rain and urban smog good chances of lasting success. Participants were less sure of an environmental market for the control of induced climate change. Participants gave reasons for their views; they were generally convinced of the market's cost-effectiveness compared with CAC control. But they had concerns about complications such as transaction costs. The participants' reasoned support of this regulatory innovation, in the light of complex evidence and after serving in a kind of jury over two years, makes for an appropriate conclusion to this part.

REFERENCES

Burtraw, Dallas. 1995. "Cost Savings Sans Allowance Trades? Evaluating the SO_2 Emission Trading Program to Date." Discussion Paper 95-30. Washington, DC: Resources for the Future. September.

Fishkin, James. 1996. *The Voice of the People.* New Haven, CT: Yale University Press.

Joskow, Paul L., and Richard Schmalensee. 1996. "The Political Economy of Market-Based Environmental Policy: The U.S. Acid Rain Program." Center for Energy and Environmental Policy Research, Massachusetts Institute of Technology, Cambridge, MA. March.

Montgomery, W. David. 1972. "Markets in Licenses and Efficient Pollution Control Programs." *Journal of Economic Theory* 5(3): 395–418.

Prager, Michael A., Thomas Klier, and Richard Matoon. 1996. "A Mixed Bag: Assessment of Market Performance and Firm Trading Behavior in the NO_x RECLAIM Program." Federal Reserve Bank of Chicago. Working Paper Series 1996/12, Chicago, Illinois. August.

South Coast Air Quality Management District (SCAQMD). 1996a. "First Annual RECLAIM Program Audit Report." Report 91765-4182. Diamond Bar. SCAQMD CA. January 12.

South Coast Air Quality Management District (SCAQMD). 1996b. "Intercredit Trading Study." SCAQMD Report 91765-4182. Diamond Bar. CA. January.

Tietenberg, T. H. 1985. *Emissions Trading, An Exercise in Reforming Pollution Policy.* Washington, DC: Resources for the Future.

[3]The methodology and objectives of the deliberative opinion poll are described in Fishkin (1996).

2.2

❑

DEVELOPMENT OF AN EMISSIONS REDUCTION MARKET SYSTEM FOR NORTHEASTERN ILLINOIS

Roger A. Kanerva and Richard F. Kosobud

INTRODUCTION

❑ Ozone levels in many U.S. big-city regions remain a stubborn and serious air quality problem. While the progress obtained in reducing lead and carbon monoxide concentrations has been significant, the progress in reducing ozone levels, as the hot summer of 1995 revealed painfully in the Chicago region and elsewhere, has not been sufficient to meet health standards. In order to more fully achieve ozone reduction goals, the Illinois Environmental Protection Agency (IEPA) proposes to utilize a new incentive-based program that will complement current controls. This paper describes the program's objectives and main features as it gets under way.

Reducing ozone concentrations is worth our best efforts. Ozone at high levels reduces the ability of normal lungs to function and does greater damage to the lungs of the young, the old, and the asthmatic. The reduction in lung performance can reverse itself, but continued exposure at high levels of concentration can result in more permanent impairment. Ozone also brings about economic and ecological losses by damaging agricultural crops, trees, and ecosystems.

The magnitude of the task is indicated by table 2.2.1, which traces ozone exceedances for the area's severe nonattainment area. Exceedances are readings of 125 parts per billion (ppb) by volume or higher, the current health standard set by the federal government, as detected by one or more of the 21 monitor stations strategically

61

Table 2.2.1. Summary of Ozone Exceedances for Illinois, Indiana, Michigan, and Wisconsin: April, May, September, October—1988 Through Mid-September 1995

Date	State	Site	Value (ppb)
April 25, 1990[a]	MI	Muskegon	130
May 13, 1991	IL, WI	Waukegan (IL)	126
		Chiwaukee (WI)	146
May 15, 1991	WI	Chiwaukee	170
May 22, 1991	MI	Muskegon	129
May 28, 1988	WI	6 Sites	149 high
May 28, 1991	WI	6 Sites	176
May 29, 1988	WI	4 Sites	140 high
May 29, 1991	MI, WI	Muskegon (MI)	147
		Manitowoc (WI)	130
May 30, 1988	WI	6 Sites	154
May 30, 1991	WI	Chiwaukee	143
May 30, 1994	WI	Bayside	128
May 31, 1988	IL, WI	Chicago	133
		Evanston	138
		Waukegan	135
		6 Sites (WI)	150 high
May 31, 1991	WI	Chiwaukee	132
Sept. 3, 1988	WI	South Milwaukee	138
		Racine	127
Sept. 4, 1990	IL	Nilwood	128
Sept. 5, 1995	WI	Chiwaukee	126
Sept. 6, 1990	MI	Georgetown	138
Sept. 7, 1991	WI	4 Sites	133 high
Sept. 13, 1990	WI	Newport Beach	126

Official Ozone Seasons by State	
Illinois	April–October
Indiana	April–September
Michigan	April–September
Wisconsin	April 15–October 15

[a]Data may be invalid or nonrepresentative. Although the temperature on that date was 87°, it was windy with considerable cloud cover, and thunderstorms were in the vicinity. These conditions are not particularly conducive to ozone formation. Furthermore, surrounding monitors, as well as monitors throughout the region, showed readings of from 80–90 ppb. The monitor recording the high reading had been out of service for the 4–5 hours directly preceding the reported exceedance. The combination of these factors strongly indicates that the April 25, 1990, exceedance in Michigan was the result of a faulty reading.

located in the area. The severe nonattainment area covers six counties and two townships of northeastern Illinois, six counties of Wisconsin, and two counties of Indiana, all of which have recorded high levels in the past. The exceedances measure the unfinished business of ozone control. Fundamentally, that unfinished business is finding cost-effective ways to reduce the emissions of ozone precursors, volatile organic materials (VOM) and nitrogen oxides (NO_x), so that the atmospheric concentrations of these precursors are held to acceptable levels.

Reducing ozone concentrations was one of the goals of the national Clean Air Act of 1970. Progress during the next 20 years being less than satisfactory using traditional regulatory measures, Congress addressed the problem again in the major amendments of 1990 in which, for the first time, market-based approaches were explicitly mandated for control of acid rain and indirectly allowed as an option for states to control ozone.

To take advantage of this new tool for reducing the region's ozone, the IEPA moved quickly in 1991 to develop a strategic planning process that gave priority to using a market-based approach. There were few precedents for the administrative work ahead. It meant designing a comprehensive market system that could account for the special features and science of ozone formation. It meant securing the views and cooperation of all affected parties, including business and environmental groups. And, above all, it meant providing full information about the goals and workings of the system so that public confidence and support could be obtained. This paper describes the emissions reduction market system that emerged from this process and is intended to be a contribution to the understanding of that system and to its ultimate successful realization.

DEVELOPMENT PROCESS FOR AN EMISSIONS REDUCTION MARKET SYSTEM (ERMS)

There were a number of reasons why the IEPA moved to implement a major new regulatory approach. Prior to 1990, clean air legislation typically required specific standards or technologies for control of the rate of pollutant emissions. Regulatory agencies were absorbed in monitoring and enforcing these detailed prescriptions, which were often applied uniformly and without regard to cost to all sources and which justifiably earned the sobriquet of *command-and-control regu-*

lation. Although the rate of emissions was regulated, total emissions were not. Total emissions were determined by nonregulated emission features such as the number of sources and their use over time. Both the regulating and regulated communities came to realize, as did many concerned observers, that further tightening of traditional regulation could not reach environmental goals without encountering rapidly increasing control costs and mounting regulatory effort. Neither environmental ends nor the economic vitality of the region would be well served by this course of action, as was recognized by local and state governments. As an alternative, an emissions reduction market system had attractive features and could be designed to complement, not replace, existing regulation.

The market system could provide for cost-effective control of ozone if those sources with the cheapest emission reduction costs could be led to carry out the deepest cuts. Based on its experience with traditional permit regulation, the IEPA knew that ozone-causing emissions came from numerous sources of varying size. Significant emission control cost variations among these sources create the potential for large savings. If emitters with high reduction costs were given the discretion and incentive lacking in traditional regulation to transfer their cuts to emitters with low costs, both could benefit financially. Moreover, the public interest would be served by the savings in resources used in cleaning the air, which could then be spent in other important areas.

A market system also creates an incentive for emitters to devise cheaper new practices and future technologies for reducing emissions because the reductions so obtained yield credits that can be sold. This becomes another major source of savings, a source lacking in traditional regulatory approaches. These new measures should lead to more pollution prevention efforts to reduce emissions at all stages of the production process, including product reformulation and redesign. In sum, the market system allows emitters, with their detailed, on-the-spot knowledge, to reduce emissions in the most cost-effective manner possible; and it allows governments, with their responsibility for the public interest, to set the goals and to monitor and enforce the market rules.

A prefeasibility study by the IEPA made clear these potential advantages (Palmer Bellevue 1992). Ozone formation presents challenges, however, to the design of a workable market. Can a market be designed to cover the varied and variously located emission sources: stationary, mobile, and area? Can trades of emission allowances issued for a season effectively reduce episodic ozone exceedances? Can the market adjust to policy changes as they reflect the latest scientific findings of the ways in which NO_x and VOM combine with hot weather to form ozone? Can a seemingly anonymous market with autonomous buyers and sellers secure the support of participants and the public as an environmental control mechanism?

A more intensive feasibility study was undertaken to address these and related questions, to exchange information on preliminary market designs with all interested groups, and to review the limited experience elsewhere with market systems.

The Los Angeles Regional Clean Air Incentives Market (RE-CLAIM), then just emerging from the developmental stage, provided helpful information on such issues as banking and initial allocation of emissions. Important differences exist between the regions, however, with each having unique geographical, economic, meteorological, and emission characteristics, so that no simple carryover of market designs was possible.

The sulfur dioxide (SO_2) allowance trading program to control acid rain specified in the Clean Air Act Amendments of 1990 (CAAA '90) also provided useful guidance on the emissions contract, auction design, and enforcement procedures although, again, important differences had to be recognized. Sulfur dioxide emission sources are few in number, large in scale of emissions, and mainly fixed in location. Emissions come mainly out of smokestacks that are easily monitored compared with the numerous, sometimes fugitive outlets for ozone precursor emissions. Acid rain is considered neither to be a seasonal problem nor to have small-area impacts. Again, no simple copying of a market design was possible.

The truth of the matter is that, while the economic theory of the gains from emissions trading had been analyzed in depth and while limited exchange of emission reductions had been attempted by permitting sources to purchase offset credits or to trade under a bubble or to net increases and decreases within an enterprise, the design of a comprehensive workable market for ozone control was very much in the early stages of development.

It was considered essential that the IEPA conduct extensive outreach efforts during this feasibility study so that representatives from business, from environmental and other public interest groups, from involved governmental agencies, and from academia could make their comments on the evolving market design. Follow-up meetings addressed specific features such as baseline emissions, staged reductions in emissions, and banking proposals. The IEPA's endeavor to secure cooperation and agreement rather than confrontation in this new enterprise was to carry over to every phase of the design process. Based on this outreach work, a report was issued concluding that the approach was both cost-effective and doable (IEPA 1993a).

In what was envisioned to be a final phase, an emission market design team was assembled to focus on the control and trading of NO_x emissions. This was considered a logical place to start because the major sources for this ozone precursor were fewer in number, larger in size, and mainly fixed (or stationary) in location. At a later date, the market could be extended to cover VOM precursors. Volatile organic material emission has a more complicated mix of stationary, mobile, and area sources. It should be noted that, because of constraints from traditional regulation, such as standards for vehicle emissions, vehicle inspection and maintenance procedures, reformulated gasoline, and requirements for reasonably available control technology (RACT), VOM emissions were assumed, incorrectly as it turned out, to be more manageable.

The first market design based on this perspective was presented in a report by Mary Gade, director of the IEPA, at a conference held in Chicago in June 1993 and published by the Federal Reserve Bank of Chicago as one paper in a series (Kosobud et al., eds., 1993). A more detailed "Draft Proposal: Design for NO_x Emissions Trading" was subsequently released in September, 1993, for public review and comment (IEPA 1993b). Unexpectedly, however, new scientific findings on ozone presented the IEPA and the design team with a dramatic and demanding new assignment.

The Lake Michigan Ozone Study, sponsored by the states in the nonattainment area, had developed an advanced emissions-meteorological model to simulate ozone formation under varying circumstances and policy options. Utilizing the latest data to calibrate the model, the scientists found that reducing NO_x emissions alone actually increased ozone formation in parts of the area. Nitrogen oxides have a twofold impact on ozone, acting both as a cleansing agent removing ozone and as a precursor agent contributing to its formation. The key appears to be the ratio of NO_x to VOM in the atmosphere. In the Chicago area, reducing the former but not the latter lessens the cleansing action more than the formation action. The scientific conclusion was clear for the region: VOM emissions required first priority for controlling ozone.

Another key finding was that the concentrations of NO_x, VOM, and ozone being transported into the nonattainment region were very high. Emissions outside the nonattainment area, even outside the state, were contributing to the local ozone problem and would, if left unattended, require much larger reductions of emissions within the nonattainment area than had previously been estimated.

To confront the first problem, the IEPA, in the spring of 1994, acknowledged the significance of the scientific evidence by shifting gears and asking the design team to move ahead on a VOM trading system. While the emission pattern of VOM was more complicated than that of NO_x, there appeared to be many opportunities for cost-effective trading that could be exploited by a market system.

To confront the second problem, the incoming emissions or boundary issue, the agency assumed a leading role in pointing out that this was a question that required national attention. Mary Gade was subsequently appointed by the U.S. EPA to chair an advisory group, the Ozone Transport Assessment Group, to make recommendations regarding national inventories of precursors and expanded regional transport modeling and to suggest national, regional, and local control strategies. One option, for example, was to reduce NO_x emissions in these outlying areas while the local environmental market reduced VOM emissions.

The Illinois VOM market design team was expanded to reflect the wider range of interests that would be affected by a redesigned market. Its composition included representatives from the Environmental Defense Fund, affected businesses, and consultants—evidence of the agency's continuing determination to secure the collaboration of a cross section of informed and concerned groups. Further expert advice was secured from a formal peer review group whose members, while not participating in the design, could provide an objective assessment of, and suggestions for, amending market proposals. They emphasized, for example, the importance and desirability of keeping the system as simple as possible to ensure its workability and support. Members of the design team and peer review group are listed at the end of the study.

The first product of the expanded trading design team was a draft proposal on VOM trading (IEPA 1994). Once again, the IEPA met with interested parties to exchange views on this draft proposal and to encourage the submission of written comments. The IEPA shared with these groups its estimate that the operation of this system could result in more than $160 million in savings for the participating sources. Studies in Los Angeles have shown that savings such as these could significantly cut the cost of further ozone control. The final design proposal for a VOM emission reduction market system was published in March 1995. It embodied the results of comments from the design team and concerned and affected interest groups, the findings of scientific modeling of the area, and the judgment of the IEPA. It should be noted that additional changes in the program were envisioned during the rulemaking process.

AGGREGATE EMISSIONS REDUCTION:
GENERAL CONSIDERATIONS

Establishing an environmental market enables the government to set an overall goal for emissions reduction and to decentralize the decision concerning the attainment of that reduction to the emitter, a cost-effective division of labor between regulating and regulated communities. There remains, however, the matter of the government's choosing from among the environmental market types available, an important choice since they differ with respect to performance characteristics and monitoring requirements.

The most comprehensive is the establishment of an areawide emissions budget with source allocation, or "cap-and-trade" market. This market has the advantages of encompassing all sources, thus enabling emitters to search out the most cost-effective emission reduction possibilities wherever they may be. This market also encourages future innovation in reducing emissions across the full range of sources. However, this market also has many requirements. It requires estimates of emissions from the full spectrum of sources, determination of baseline emission levels, agreement on an emissions cap and its projected rate of decline, agreement on a method of allocation of capped emissions among the sources, and effective techniques for monitoring and measuring each source's emissions. This is a formidable list of requirements, especially in the case of ozone control, which has numerous and frequently small sources.

Less demanding is the open market, as distinct from the cap-and-trade, or "closed," market, in which trading can begin operating before agreement is reached on a comprehensive cap or source allocation. The seller can receive credit for discrete quantities of emission reductions measured from its own baseline of actual or legally allowable emissions, whichever is lower. The buyer can use the credit instead of complying by reducing its own emissions. An open market can thus begin to realize some of the cost-effective gains of trading but, because of its partial coverage, it will perform in theory less efficiently than the cap-and-trade model.

In order to reap the advantages of the cap-and-trade model, the IEPA market design builds in as many of its features as possible. Larger point or stationary emission sources of VOM are required to be part of the market system. A provisional cap, the allocation of emission permits to individual point sources, and a projected rate of decline are established for these sources. The important mobile and area sources, for which caps and rates of decline are too numerous and complex to establish under present conditions, can be tapped, in part, by traders bringing about creditable emission reductions in a manner we shall discuss shortly.

Essential to the working of a cap-and-trade market are credible aggregate emission data. Figure 2.2.1 reveals that aggregate emissions in 1990, the base year, totaled a little over 1200 VOM tons per day (TPD) from anthropogenic sources—26% from point sources, 52% from mobile on-road and off-road sources, and 22% from area sources (in rounded percentages). To reach cleaner air attainment goals in the year 2007, total area emissions must be substantially reduced. Precisely how much they are to be reduced depends on several uncertain future developments, among them the success of reducing the amount of precursor and ozone concentrations that will come into the area from outside sources. Because this amount will depend on national and regional policies beyond the control of the IEPA, it is necessary to develop several scenarios for reducing local emissions conditional on these policies and their implementation. These scenarios are detailed in table 2.2.2.

1990 CHICAGO INVENTORY SUMMARY
VOM EMISSIONS

Nonattainment Area Anthropogenic Sources

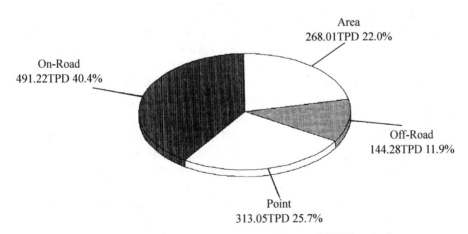

Figure 2.2.1. 1990 Chicago inventory summary of VOM emissions.

Table 2.2.2. Chicago attainment strategy

	Emissions of VOM (Tons/Day)[a]					
	(1)	(2)	(3)	(4)	(5)	(6)
	Year					
Emission category	1970	1990	1996[b]	2007[c]	Target[d] (50%)	Target[e] (60%)
Point	340	313	206	227	164	104
Area	549	268	216	175	175	161
On-highway	992	491	257	155	155	150
Off-highway	109	144	145	106	106	65
Totals	1990	1216	824	663	600	480

[a]Estimates represent total anthropogenic volatile organic material (VOM) emissions.
[b]Year 1996 control measures.
[c]Year 2007 control measures.
[d]To meet 50% reduction level within the Chicago NAA by the year 2007.
[e]To meet 60% reduction level within the Chicago NAA by the year 2007.

Table 2.2.2 shows that total emissions from all sources are expected to be reduced by 1996 to a little over 800 tons per day. These reductions will be achieved by traditional control measures before the market system gets into operation. The scenario of column 4 assumes that, again before the market system gets into operation, the continuance of traditional controls, now in effect or scheduled to be effectuated, will reduce total emissions to 663 tons per day in 2007, insufficient for National Ambient Air Quality Standard (NAAQS) ozone attainment. Note, however, in the nonpolicy scenario, that point source emissions, solely under traditional controls, would be expected to increase by 10% between 1996 and 2007.

Attainment requires, in the two policy scenarios shown in columns 5 and 6, total emissions of 600 and 480 tons, respectively, dependent on reductions of precursor emissions outside the nonattainment region. In these illustrative policy or market scenarios, the emission reductions market system starts up in 1999 to make its contribution to future reductions. These two scenarios differ by the amount of outside generated precursor and ozone concentrations entering the local area. In the scenario of column 5, a 50% reduction in total area emissions from 1990 is required for attainment based on significant reductions of incoming concentrations. The point source share of this reduction can be achieved by cutting allotments of ATUs so that emissions are reduced to 164 tons per day by 2007, a reduction of about 48% from 1990. This means that the ERMS scheme, fully operative in 1999, will require annual cuts in point source allotments of about 6% per year from 1999 to 2007.

In the scenario of column 6, less significant cuts of incoming concentrations are achieved so that a heavier 60% reduction in total area emissions from 1990 is required. However, not all extra reductions need to come from point sources. Point sources or other traders may find cost-effective ways to reduce area and mobile emissions for which they would receive credit. (Buying out two-stroke lawn-mowers and old-fashioned outboard motors for boats are examples of possible reduction sources.) Hence, the scenario in column 6 envisions additional reductions in area and off-highway sources. Point source emissions in this scenario are reduced to 104 tons in 2007, a reduction of about two-thirds from 1990 levels. This would imply annual rates of reduction in ERMS allotments of a little over 7%.

CONSTRUCTING AN EMISSIONS REDUCTION MARKET SYSTEM FROM SCRATCH

In ordinary shopping, we take the market for granted, pausing only in a cursory manner to look up the location of the seller, check the commodity's characteristics, survey prices of other sellers, make the decision and the payment, and take home what we have bought. Behind the scenes are such market rules as property rights, antitrust regulations, liability laws, and enforcement procedures that enable us to make these transactions with reasonable confidence. What we will describe, however, is a market for cleaner air with similarities to ordinary markets but also with striking differences. It is a market whose commodity is a right to emit, but the consequence is fewer emissions than in the no–environmental market case. It is a market that cannot be taken for granted but must be constructed piece by piece, each piece to be appraised for conformance with the public interest.

So that this construction can be properly evaluated, we provide details on its major features: the expected participants, the "commodity," the market's general working structure, the proposed timing of implementation, and the monitoring and enforcement aspects. We conclude with comments on forthcoming rulemaking.

Market Participants

The final design proposal aims for a transaction process that is user-friendly and service-oriented. The intent is to construct a flexible market that would initially cover the larger point sources of VOM emissions but would be capable of extension to include other sources. A market exchange system would be established for servicing the participants. Each source would be eligible to buy and sell trading units via transaction accounts. Each participant would designate a responsible official who would apply for an account and manage the source's transactions.

The Emissions Commodity: Allotment Trading Units

Each participating source would be issued allotment trading units (ATUs) for its seasonal VOM emissions. Each ATU would represent 200 pounds of emissions and be dated for a two-year life, which means it could be banked for one year.Allotment trading units can be bought, sold, or used up in emissions. The number actually available to a source for banking or trading would be determined by subtracting the amount of each seasonal allotment needed to account for the source's actual emissions during the season from the source's total allotment. Consequently, the cost-minimizing enterprise has an incentive to compare the marginal cost of reducing VOM emissions by 200 pounds with using an allocated ATU with a discoverable market price. If the control cost is less, the enterprise can sell or bank the ATU; if the ATU price is less, the enterprise may choose to emit and turn over its ATU. The potential for saving control costs is clear.

Each ATU would have an identification number for tracking purposes. IEPA would maintain an ownership database containing the identification numbers of all ATUs and a market exchange electronic bulletin board. As part of the flexible market system, the IEPA would also issue ATUs to other authorized enterprises or persons who have generated valid emissions reductions in a manner to be explained shortly.

Emissions Banking

Sources are autonomous in determining how their allotments are going to be used. Any allotment trading units not needed by a source for its own emissions that season could be traded to another party or banked for use in the next ozone season. Note that banked ATUs serve to clean the present air more rapidly. Banking may be desired by a source to avoid a crunch at the end of the record period when ATUs must be given up for emissions. Such banking should also minimize ATU price spikes due to many sources falling short of ATUs at the same time. Banking enables sources to smooth out cyclical swings in production and, hence, emissions. Limiting banking to two years should avoid the problem of emission spikes due to many sources using substantial amounts of banked ATUs at the same time.

Transaction Process and Compliance Accounting

The minimum trading amount would be one ATU. Valid trades would involve the transfer of ATUs between active accounts by authorized persons. The market system database would keep records of transactions, banking, and usage of ATUs for each source.

A 90-day reconciliation period would follow each 5-month ozone season. During this period, sources would be given time to compile accurate results for actual VOM emitted. This period would also provide time for participating sources to obtain ATUs needed to cover any excess emissions. Upon use for compliance, an ATU and its corresponding number would be retired. At the end of the reconciliation period, the IEPA would do a computer check of actual emissions reported versus ATUs held by each source. Discrepancies would be cause for taking enforcement action.

Market Assurance

Concerns have been expressed about possible monopolization of ATUs by one or a few sources. Questions have also been raised about how enterprises new to the region's economy could obtain ATUs to cover their emissions. To address these concerns and questions, the IEPA would establish a special ATU set-aside, called the alternative compliance market account (ACMA), by deducting 1% from each source's initial seasonal allotment. These reserved ATUs would be available to sources needing ATUs and unable to obtain them elsewhere. Under certain conditions, the ACMA may be drawn down to a negative balance by issuing ATUs in anticipation of future emissions reductions. If such reductions are not obtained, however, IEPA can deduct what is needed from the allotments of the participating sources and compensate them accordingly.

Intersector, Interpollutant, and Interregional Trading

At any time, market participants may wish to generate ATUs by devising ways to reduce area or mobile source emissions, a transaction the IEPA would encourage. For example, high-emitting vehicle or outboard motor scrappage programs may be used to generate cost-effective emissions reductions among mobile sources, thereby entitling point sources to receive an allotment of ATUs for such action. Scrapping high-emitting lawnmowers is another example that could generate cost-effective reductions among area sources. There are probably many more ways to broaden the market that clever traders can search out. The region benefits in that air is cleaned more cheaply as more sources are covered. In these instances, the current and future allocations of the ATUS would depend on the nature of the emissions reductions.

Extending the market-based approach outside the nonattainment area would be another important step in bringing the movement of precursors and ozone from outlying areas under control. In these outlying areas, reducing NO_x emissions would be effective in limiting ozone formation, as modeling simulations have shown. A market system for large NO_x emitters in Illinois could be both feasible and cost-effective. Extending this market beyond state boundaries holds out the promise of additional control and further savings.

In sum, success in the implementation of the market design for control of VOM point source emissions should be an impetus to these market extensions. This success is unlikely to be achieved overnight. Constructing an effective new market requires implementation to be carried out in a timely and reasonable way to secure the full participation of traders and the support of concerned observers.

Moving to a Seasonal Market system

It must be emphasized that the market system would not replace, but would be built on and complement, the existing traditional approach to source control. During the period 1994 to 1996, the required 15% reduction in VOM emissions is to be obtained through traditional year-round control, mainly by means of reformulated gasoline and RACT controls. These traditional controls and permitting programs would continue, but the emissions reduction market would become the main instrument for securing further reductions from stationary sources during 1997 and 1998 and after 1999. Furthermore, it will be recalled that the market would be in operation only during the seasonal control period, which would extend from May 1 to September 30, coinciding with those periods when actual exceedances of the ozone standard have been recorded in the air shed.

Baseline, Cap, and Rate of Decline: The Period 1999–2002

By January 1, 1998, point sources with 10 tons or more of seasonal emissions during 1997 and the baseline period would submit an ERMS application containing VOM emissions data as specified. The initial seasonal allotments of ATUs would be issued for 1999. Current estimates reveal that 283 sources, contributing about 90% of relevant point source emissions, would be subject to this reporting. Any source that newly generates 10 tons or more of seasonal emissions after 1999 would be required to come under the market rules. Other sources, typically smaller emitters, may participate in the ERMS by teaming with a participating source and submitting a proposal for reductions by the smaller source in seasonal VOM emissions. Once approved, these reductions would then become additional ATUs available to the larger source that would pay for them.

Actual VOM emissions data in tons per season would be used to determine the initial seasonal allotment for each source. A source could select from its own emissions history the average of the two seasonal allotment periods with the highest VOM emissions during 1994, 1995, or 1996. An average of these years becomes the benchmark from which reductions are calculated. If emissions during 1994, 1995, or 1996 were nonrepresentative of the source's emissions, the source may substitute for one or two of these years from the emissions recorded during 1992, 1993, or 1997.

For 1999 and thereafter, all 283 sources emitting 10 tons or more would be operating under seasonal emissions allotments. Each source would receive an initial allotment in 1999 determined by the baseline emissions for that source, subject to a 12% reduction that has three components: a 9% reduction to account for reasonable further progress, a 1% reduction for the ACMA, and a 2% contingency reduction. The need for any further reductions will be determined by the IEPA based on the results of the regional NO_x control evaluation.

Exclusion from reductions will be determined by the regulatory agency upon demonstration that the emission unit is achieving the best available technology. Also, any participating source that is permitted to operate during start-up, malfunction, or breakdown in

accordance with the Illinois Administrative Guide will not be required to reduce VOM emissions during the specified time interval.

The Period 2000–2002

The Illinois EPA may file with the Illinois Pollution Control
Board a written petition to continue downward adjustment of allotments of ATUs based on the determination that additional emission
reductions are necessary to fulfill the requirements of the Clean Air
Act. If the determination is not made, ATUs will be issued each year
at the same level.

Monitoring

Accurate, reliable, and consistent quantification of emissions is a
fundamental requirement for an effective market system, and for
monitoring and enforcement of its rules. For emissions of VOM, this
poses a number of challenges given the number and diversity of
sources. It is neither feasible nor realistic to require all sources to
have continuous emissions monitoring, as is required in the SO_2
allowance program. There are well-established methods and techniques that have been and can be used to quantify VOM emissions.
Furthermore, the market system can be expected to stimulate further
progress in quantification technology and methods.

Several principles of quantification can serve as foundations for
the monitoring and enforcement provisions of the market system.
Traditional permitting regulations provide, in many instances, relevant data although measurement of aggregate emissions may require
more sophisticated techniques. The emissions generated by a unit
can generally be determined by a material balance calculated on the
VOM contained in raw materials and/or on a given relationship to
particular material inputs or operating data and/or on a fixed relationship to hours of operation. Measurement techniques should
assure, with a high degree of confidence, that VOM emissions are at
or below a given level. If a source cannot be quantified, it should be
excluded from market transactions.

Enforcement

The agency would pursue statutory penalties for violations arising
from any of the following four situations: (1) late reporting of

seasonal emissions, (2) inaccuracies in emission data or in reports submitted by a participating source, (3) late filing of an application by a participating source, and (4) actual noncompliance by a participating source with emission rate limitations and other regulatory requirements.

An emission excursion or exceedence compensation would be applicable to participants whose actual seasonal emissions exceed a seasonal allotment and for which there are not sufficient banked amounts available or trading units purchased on the open market during the reconciliation period. For a first occurrence, a participant would be required to purchase replacement allotment trading units from ACMA at a penalty price of $1,200 per ATU, which is $200 above the set price. For a second occurrence, the participant would be required to pay $1,500 per ATU. If the ACMA set-aside is insufficient, the required deductions would be made from a source's subsequent allocations of ATUs.

More compliance reporting may be required after a second excess emissions excursion takes place. In this case, the IEPA could require the violating source to begin monthly reporting of VOM emissions. This would give the IEPA much more timely information about the pattern of emissions related to a problem source. IEPA would also require a violating source with an allocation to submit an excursion management report detailing the cause(s) of the excess emissions and the steps the violator would take to prevent future emissions excursions.

UNIQUE ASPECTS OF VOMS: ISSUES AND CHOICES

Not all VOM emissions are alike: They differ in toxicity, in ozone formation reactivity, and in geographic contribution to ozone formation These aspects require careful consideration for their potential impacts on the market design.

In general, the pursuit of VOM emissions reductions should facilitate air toxics control under Title III of the CAAA. The incentives associated with emissions trading should serve to get toxic emissions reductions at a faster or greater pace than command-and-control regulation would. That is, both ozone—a criterion pollutant dangerous only at high concentrations, which falls within Title I—and benzene—one of the hydrocarbon precursors of ozone and a toxic substance dangerous at low levels, which falls under Title

III—will jointly be reduced under the market system, with additional protection provided for the toxic substances. To ensure consistency with toxic control by maximum achievable control technology (MACT) required under Title III, the specified approach for the market system is not to allow participating sources to avoid MACT reductions through emissions trading nor to distinguish VOM emission reductions required by MACT from other reductions. This approach is consistent with the way MACT has already been handled for the purpose of ozone attainment. By adopting this approach, sources of air toxics would not be excluded as players in the ozone attainment program.

Varying reactivity of different VOM species in forming ozone—some are more potent in this regard than others—is not specifically considered in the emission reductions required by the CAAA. Existing control measures that apply to specific categories of equipment also do not address VOM reactivity. Volatile organic material species tend to be thoroughly mixed at most emission sources. At this point, there is no reason to believe that an extensive pattern of adverse impacts would ensue from nondifferentiated VOM trading units. The IEPA will continue to evaluate this factor using real data from an operating emissions trading system.

If ozone concentrations were uniform over the region and emissions everywhere contributed to that ozone to the same extent, no distinction would then need to be made in the market among ATUs from different locations. Although that uniformity is not the case, the present market design has proceeded on that basis as it greatly simplifies the transaction and recordkeeping process. This is not to deny that the ozone plume has a spatial dimension and that ozone in heavily populated areas is more damaging than elsewhere. Nor is it to be denied that emissions from source A may contribute more to ozone formation than emissions from source B, so that trades from B to A may seem to go in the wrong direction. One major distortion in this case would be for a rural source B to sell emission credits to an urban source A, thus transferring emissions to more populated areas.

The offsetting considerations to these varying geographic attributes of ozone is that the location of the plume is subject to changing wind direction, leaving its exact location uncertain. Ozone frequently forms over Lake Michigan when stagnant hot air and precursors encounter cold water. The ultimate destination of the plume is subject to the vagaries of the wind. The uncertainties are such that the gains in assigning varying weights to ATUs depending on their location would probably be more than offset by limitations in the efficiency of the market. It is a matter for continuing reexamination and for close future monitoring, using the more than 20 ozone receptors scattered about the Chicago region.

STATE LEGISLATION AUTHORIZING A MARKET SYSTEM

The IEPA wanted to ensure that there was a clear mandate to proceed with regulatory development of the emissions reduction market system. Thus, legislation was introduced in the spring, 1995, session of the Illinois General Assembly. Senate Bill 460 generated another round of dialogue about how best to achieve the requisite VOM emissions reductions and brought renewed attention to the policy considerations and structural components of the market system. Some concerns that had developed in the business community were addressed by changes in the market design that aimed to achieve a balance of interests.

The greatest concern of business involved a basic uncertainty about the eventual economic and technical limitations that a participating source might encounter. In essence, the business sector wanted a safety net to rely on if extreme economic consequences arose in the operation of this VOM emissions market. The design team had considered this concern previously and had proposed an innovative solution: the alternative compliance market account (ACMA). This concept involved establishing an account of last resort that would be operated by the IEPA and would be accessible to certain sources that have no other viable compliance options. Allotment trading units could be purchased from this account based on a preset rate that reflects the upper range of known VOM control costs. IEPA would, in turn, use the funds collected to generate equivalent emissions reductions from within the nonattainment area to balance out the system. In this manner, the IEPA would provide a service for sources that run out of affordable ways to achieve compliance. Both business and environmental interests accepted this concept as a good way to address this issue.

Another major concern involved the protocol for setting emissions baselines that would define the starting point for each source's allotment. Business interests wanted more flexibility to account for unusual conditions at a facility. They also did not want to be penalized, so to speak, for certain early reductions by being given a smaller seasonal allotment. Other interests raised a concern about strategic behavior regarding the baseline if the determination involved years that were too far in the future. A source might elect to increase actual emissions so as to get a larger allotment. Ultimately, the IEPA agreed to address these concerns in a cooperative manner as part of the rule-making process.

The final major concern involved enforcement provisions. Business interests took the position that the design proposal created a sort of triple jeopardy: A source would be potentially subject to traditional enforcement, to emissions excursion compensation, and to noncompliance fees. It was agreed that the first two aspects would be deleted from the bill and that more work would go into design of a workable set of enforcement provisions during the rule-making period.

Senate Bill 460 eventually received unanimous support from all affected interests and no dissenting votes from the legislature. Indicative of the legislative support was the comment of the chairman of the Senate Environment and Energy Committee in his opening remarks at a hearing: "S.B. 460 is the best legislative action by the IEPA in a long time."

CONCLUSIONS

An emissions reduction market system has attractive features that we believe warrant the efforts expended in its development. It can provide incentives for cost-effective ways to control ozone, no small matter in an era of rising regulatory costs and limited budgets. It can stimulate new and innovative control measures in the future, something that traditional regulation often fails to do. It provides for decentralized decision making in an anonymous manner, which should make for improved relations between the regulated and regulating communities. These are powerful arguments for this approach to achieving environmental goals.

There are also features that require that we pause to set caution flags along the route ahead. Gains from such a market system have yet to be proved in practice although the first results from SO_2 allowance trading seem promising. The market institutions outlined in this study will have to be constructed in such a way as to secure the confidence of the public and yet must be simple and acceptable to traders so as not to impose heavy transaction costs. The market must be seen as a fair way to achieve cleaner air and must, at the same time, work in an efficient manner.

It will take time for full implementation of all the market features outlined in this study, some of which will be altered by experience. This study has presented the construction design and details as an aid to future appraisals. A successful market performing in the public and private interest will not just happen automatically but will take contributions from all concerned parties. The IEPA is fully dedicated to making its contribution.

REFERENCES

Clinton, Bill, and A. Gore. 1995. "Reinventing Environmental Regulations." Unprocessed. White House, Washington, DC. March 16.

IEPA (Illinois Environmental Protection Agency). 1992. "Four Year Strategy for Environmental Progress." Springfield, IL. April.

IEPA. 1993a. "Feasibility Study for Market-Based Approaches to Clean Air." Springfield, IL. September.

IEPA. 1993b. "Draft Proposal: Design for NO_x Trading System." Springfield, IL. September.

IEPA. 1994. "Draft Proposal: Design for VOM Emissions Trading System." Springfield, IL. October.

IEPA. 1995. "Final Proposal: Design for VOM Emissions Trading System." Springfield, IL. March.

Kosobud, Richard, Donald Hanson, and William Testa (eds). 1993. *Cost Effective Control of Urban Smog.* Chicago: Federal Reserve Bank of Chicago.

Palmer Bellevue Corporation. 1992. "Emissions Reductions Trading in the Chicago Metropolitan Area: A Pre-feasibility Analysis." Chicago. May.

DESIGN TEAM MEMBERS

John Calcagni, Principal
E^3 Ventures, Inc.

J. Cale Case, Vice President
Palmer Bellevue Corporation

Bill Compton
Corporate Environmental Affairs
Caterpillar Tractor, Inc.

Daniel J. Dudek, Senior Economist
Environmental Defense Fund

Mary A. Gade, Director
Illinois Environmental Protection Agency

Joseph Goffman, Senior Attorney
Environmental Defense Fund

Robert W. Hermanson, Coordinator
Environmental Issues
Amoco Oil Company

Alan Jirik, Director
Regulatory Affairs
Corn Products

Roger A. Kanerva, Environmental Policy Advisor
Illinois Environmental Protection Agency

Gerald M. Keenan, Vice President
Palmer Bellevue Corporation

Robert H. LaPlaca, Senior Environmental Engineer
Commonwealth Edison Company

Philip R. O'Connor, Chairman and President
Palmer Bellevue Corporation

Bharat Mathur, Division Manager
Air Pollution Control
Illinois Environmental Protection Agency

Steve Zeismann
Manager of Corporate Environmental Affairs
Abbot Laboratories

PEER REVIEW TEAM

C. Boyden Gray
Wilmer, Cutler & Pickering

Robert W. Hahn
American Enterprise Institute for Public Policy Research

Richard F. Kosobud
Department of Economics
University of Illinois at Chicago

2.21

❏

DISCUSSANT

Ronald L. Burke

A CRITIQUE OF ILLINOIS' MARKET-BASED STRATEGY FOR REDUCING VOM EMISSIONS FROM POINT SOURCES IN NORTHEASTERN ILLINOIS

BACKGROUND

❏ Northeastern Illinois has an intractable air pollution problem that is most troublesome in the summer months when ozone smog levels soar. Illinois' efforts to minimize this problem have been driven by the Clean Air Act, which became law in 1970 and has since been amended several times. More than 25 years later, despite rapid growth in the practices that produce air pollution, for example, vehicle use, manufacturing, and consumer-related consumption, the Clean Air Act has resulted in considerably cleaner air in metropolitan Chicago.

Unfortunately, the act has not worked well enough nor fast enough to prevent thousands upon thousands of illnesses and premature deaths that would otherwise have been avoided. This is evidenced by the region's failure to meet the federal ozone air quality standard, even though that standard is inadequate.[1]

---

[1]The American Lung Association sued the U.S. EPA to force a review of the ozone standard. In March, 1995, U.S. EPA's Clean Air Act Advisory Committee endorsed a draft document that recommends a 6- to 8-hour ozone standard of 0.07 to 0.09 parts per million (ppm). Such a longer-term standard would probably supplement the current 0.12-ppm 1-hour standard or a revised 1-hour standard.

Furthermore, current projections indicate that improvements from existing and planned control measures will not keep pace with emissions increases. For example, volatile organic material (VOM) emissions from point sources, which have been steadily declining since 1970, are projected to increase between 1996 and 2007 (northeastern Illinois' Clean Air Act deadline for achieving the ozone standard) despite tighter Clean Air Act standards. And, after at least 40 continuous years of improvement, mobile source emissions are expected to bottom out and then increase around 2010, when cleaner cars and fuels can no longer offset the astronomical growth in vehicle use.

Therefore, while air quality in metropolitan Chicago and other urban areas needs to improve considerably, emissions increases caused by growth in polluting activities threaten to do the opposite by offsetting the benefits of traditional pollution control measures, What's needed are new tools that enable air quality officials to assemble better plans, and Illinois' VOM cap-and-trade proposal is a good example.

ILLINOIS' MARKET-BASED APPROACH
FOR CLEARING THE AIR

Market-based programs represent promising supplements to traditional command-and-control-based air quality plans. Transportation pricing measures, for example, can minimize vehicle use and emissions by increasing the cost of driving consistent with associated air pollution- and congestion-related expenses. The additional revenues can be used to fund alternatives to single-occupant vehicle trips. The "revenue-neutral" approach to transportation pricing increases up-front driving costs (e.g., parking fees) while reducing hidden costs (e.g., property and sales taxes). As a result, the average motorist drives less because it is economically advantageous but doesn't pay more taxes or fees.

Illinois' VOM emissions cap-and-trade proposal is different in that it attempts to use a market mechanism to create a more cost-effective regulatory approach. The proposal would effectively apply a declining emissions cap against the sum of emissions from at least 283 large point sources in the Chicago nonattainment area. Instead of requiring specific emissions limits at each site, a market would be established for emissions credits that allows the sources to collectively meet a regional emissions cap. Sources that can cost-effectively reduce emissions would do so, while those that cannot would buy credits. Moreover, the monetary value of the credits creates an additional incentive for sources to reduce emissions.

This concept makes a world of sense; however, implementing the concept is the hard part. We think the Illinois Environmental Protection Agency (IEPA) has proposed a good approach, although significant flaws need to be resolved. The single most important provision of the proposed cap-and-trade program is the declining emissions cap. With the cap, the program can substantially reduce emissions and contribute to attainment. Without the cap, emissions reductions are likely to be negligible and highly unpredictable.

Other critical program components that are well designed include the provisions for the two-year maximum lifetime for emissions credits and IEPA's set-aside of credits to ensure a sufficient supply.

Nonetheless, the final proposal is far from perfect. Summarized below are some of the most important deficiencies and flaws that need to be remedied.

PROBLEMS WITH ILLINOIS' PROPOSED CAP-AND-TRADE PROGRAM

Emissions Reduction Potential

IEPA believes that the cap-and-trade program, plus the measures contained in its plan to reduce VOM emissions 15% by 1996, plus upwind emissions reductions, will allow the region to attain the current ozone standard. However, under this scenario, the agency is counting on roughly 40 to 120 tons per day of VOM emissions reductions from the cap-and-trade program, which is 19% and 56%, respectively, of the total estimated point source emissions in 1996.

To the best of our knowledge, the agency has provided no evidence that this range of reductions is feasible. If it is not, the state's attainment plan is deficient. Moreover, if the estimates are feasible but nonetheless overly ambitious, the program could fall victim to political opposition from disgruntled industries. In fact, the Chicagoland Chamber of Commerce announced its opposition to the cap-and-trade program in July 1996, shortly before IEPA planned to formally introduce the proposal at the Pollution Control Board. IEPA should demonstrate that the proposed VOM cap-and-trade program can reasonably deliver emissions reductions that are consistent with the attainment plan. If this is not possible, it will be necessary for IEPA to supplement the attainment plan with additional measures.

Exemptions

The proposal exempts emissions units that already achieve the "maximum degree of reduction ... that is currently available in practice for any similar unit." [Section 205.405(a)(3)(C) third draft of the ERMS rules.] This loophole could result in numerous exclusions that would undermine the proposed cap-and-trade program.

First, exempting sources would minimize the "technology-forcing" component of the program that is necessary to achieve the level of reductions that Illinois is counting on. Second, exemptions minimize potential emissions reductions simply because there would be fewer participating sources. Third, with a maximum number of 283 affected sources, the proposed market mechanism can ill afford even fewer participants if it is to be sufficiently robust. And, finally, because the proposal would apply a cap only on emissions rates (and not on total emissions) at exempted sources, total emissions at these sources could actually increase if production goes up.

Illinois should rethink the proposed exemption provisions. At the very least, a seasonal emissions limit should be established for exempted emissions units, and the attainment plan should be revised to reflect the reduced emissions reduction potential of the cap-and-trade program.

Predictability of Emissions Reductions

Assuming that the agency has accurately estimated the emissions reduction potential for the VOM cap-and-trade program, it still may be impossible to predict emissions levels for any one year. This is because actual emissions can vary depending on the number of banked credits that are used. The more sources that utilize credits banked from the previous ozone season, the higher the emissions will be during the current ozone season. Thus, IEPA can predict only a range of emissions for any given year, and it is imperative that the most conservative estimate be used when developing plans to meet Clean Air Act milestones and to demonstrate attainment.

Pre-MACT Air Toxin Increases

Because the VOM cap-and-trade proposal does not distinguish between toxic and nontoxic VOM emissions, a source can purchase credits generated by nontoxic VOM emissions reductions to increase toxic VOM emissions. Although increases in toxic VOMs would be limited by the proposal's cap on total VOM emissions at 1996 levels, it is still conceivable that toxic VOM emissions could increase at individual facilities. More worrisome is the possibility that multiple industries clustered together (e.g., on the southeast side of Chicago) would each use credits to increase toxic VOMs.

The risk of localized increases in toxic VOM emissions is real, even if toxic VOM emissions could be increased only by securing toxic VOM credits. After all, toxic VOM reductions in Lake County do not clear the air of hazardous chemicals in Summit, and vice versa. To prevent local toxic VOM emissions increases, the two sources would need to be located near each other.

Assuming that the Clean Air Act survives Congressional attempts to gut or obstruct its provisions, U.S. EPA must implement standards, called maximum achievable control technologies (MACT), to control 189 federally designated hazardous air pollutants (HAPs) at industrial facilities. However, MACT won't be in place at many facilities when the cap-and-trade program is scheduled to begin. In addition, Illinois has adopted its own list of toxic air contaminants (TACs), which includes nearly all the HAPs plus more than 60 additional pollutants. Unfortunately, the state has yet to adopt TAC control standards and probably won't for some time.

In order to prevent localized increases in toxic VOM emissions, IEPA should establish an emissions cap based on actual historic emissions for HAPs and TACs until such time as control standards are adopted and being enforced.

Because the cap-and-trade program would be applicable only during the ozone season, another concern is that off-season emissions, particularly toxic emissions, will increase because (1) sources shift more production to the off-season and (2) control efficiency is reduced during the off-season. Until control standards are adopted, the proposal should include some precaution against substantial increases in HAPs and TACs during the program's off-season. The emissions cap described above would be sufficient.

New Source Review Requirements

The new source review (NSR) provisions of the Clean Air Act require certain new facilities and existing facilities that substantially increase emissions in the Chicago nonattainment area to secure 1.3 pounds of VOM offset credits for every new pound of VOM to be emitted. The goal is to allow industrial growth while still reducing source emissions.

However, the proposed cap-and-trade program would apparently eliminate the NSR offsets requirement by requiring new and major modifying sources to obtain only emissions credits equal to their seasonal emissions, as opposed to credits equivalent to 1.3 times their annual emissions.

This is a bad idea for at least three reasons. First, eliminating the offsets requirement would forgo additional emissions reductions that northeastern Illinois needs to demonstrate attainment. Second, with no offsets mechanism in place, the U.S. EPA's ability to sanction Illinois for failure to comply with the Clean Air Act would be compromised. This is because the first sanctions penalty required by federal rules is an increase in the offset ratio. And, finally, Illinois simply does not have the authority to eliminate the NSR offsets requirement, which is specifically mandated by the Clean Air Act. Illinois should ensure that 1.3:1 VOM offsets are still required during the ozone season and the rest of the year.

Compliance Assurance

The proposal relies on a fragmented system of compliance assurance that the general public, and perhaps even IEPA, would have a hard time assessing. In fact, it is not clear how the Illinois EPA will monitor performance or determine whether a source has violated its emissions cap.

A simple source-by-source compliance summary that pulls together key information from different components of the proposed program is imperative for monitoring compliance and for ensuring that the program is running properly. The compliance summary should include:

1. Actual seasonal emissions in credits and in tons of VOM
2. Credits allotted for that season
3. The difference between consumed credits and allotted credits
4. Total number of credits sold (if any)

5. Total number of credits obtained (if any) and origin of the credits
6. Credit balance = allotted credits plus obtained credits minus actual number of credits used minus sold credits
7. Violation of emissions cap? Yes or no
8. Date of violation and description of penalty
9. Status of penalty
10. Has an audit been conducted? Yes or no
11. Date of last audit
12. Noncompliance or deficiencies discovered? Yes or no and brief description
13. Corrective action plan required? Yes or no and date due to IEPA
14. Status of corrective action plan

Directionality of Trades

Roughly 80% of ozone exceedances occur on days when winds are out of the southeast, south, or southwest. The cap-and-trade program would reduce ozone comparatively less if VOM emission decreases occur disproportionately in the northern portion of the nonattainment area versus the southern portion. When modeling the program as part of the attainment plan, Illinois should not assume an even distribution of VOM emission decreases. The directionality of trades should be recorded and accounted for in the modeling. Furthermore, it may be necessary to limit the number of north-to-south trades in order to maximize air quality improvement.

CONCLUSION

Solving tough problems requires initiative and leadership. Illinois has shown both in designing a program that can help residents of northeastern Illinois breathe easier. My employer, the American Lung Association of Metropolitan Chicago, and I look forward to working with IEPA to resolve the proposal's flaws and to assist with its implementation.

2.22

❑

DISCUSSANT

Cynthia A. Faur

COMMENTS OF MINNESOTA MINING
AND MANUFACTURING COMPANY
ON THE PROPOSED EMISSION MARKET SYSTEM
FOR NORTHEASTERN ILLINOIS

❑ Minnesota Mining and Manufacturing Company (3M) was pleased to have been given the opportunity to participate in the Workshop on Market-Based Approaches to Environmental Policy and to offer its comments on the proposed Emissions Reduction Market System (ERMS) developed by the Illinois Environmental Protection Agency. Since this workshop, the proposed ERMS has been revised to reflect the comments made to the agency by 3M and others. The result of these revisions is the creation of a simpler, more user-friendly trading system.

Since 1991, 3M has operated its Bedford Park facility under a voluntary emission cap. Under this voluntary cap, 3M has been able to implement several productivity improvements at its Bedford Park plant that have resulted in additional emission reductions below the emission cap. These productivity improvements have allowed 3M not only to obtain a construction permit for the installation of a new coating line, which will have emissions far below those of the lines it replaces, but also to further reduce its capped allowable emission level in 1994 and to donate the excess emission reductions to the state of Illinois for air quality improvement and to the city of Chicago to create an emission reduction credit bank.[1]

A properly designed and administered "cap-and-allocate" compliance system can promote environmental improvements while allowing businesses to operate in a cost-effective manner. The final draft ERMS discussed in the paper, "The Development of an Emissions Reduction Market System for Northeastern Illinois," in this volume represents a positive step in the development of a market-based, cap-and-allocate system for the Chicago area.

Minnesota Mining and Manufacturing believes that the emission trading system described in the paper contains several provisions that will provide businesses with the flexibility necessary to operate under a market-based system. For example, 3M strongly supports the concept of a seasonal trading period, the establishment of an alternative compliance market account (ACMA) and the agency's position that participating sources should not be penalized for emission reductions of hazardous air pollutants (HAPS), which are also volatile organic materials (VOMs). But 3M also believes that, for the final program to be effective, it must be designed in a manner that not only maximizes the air quality benefits associated with a market-based system but also provides participating businesses with the level of certainty required to operate effectively in an ozone nonattainment area like Chicago.

To this end, 3M requested that the agency consider the following suggestions. The emission baseline should not penalize sources for early emission reductions that go beyond the requirements of the state's 15% rate of progress (RoP) plan. Minnesota Mining and Manufacturing believes that the inclusion of this suggestion in the final version of the Emission Trading System will maximize a source's ability to integrate its future regulatory requirements into its business plan and thus allow the source to operate in a cost-effective manner.

Minnesota Mining and Manufacturing also believes that the trading system should be designed in a manner that both allows sources to integrate their emission reduction requirements with their business plans and does not penalize forward-looking sources that have already reduced emissions beyond future regulatory require-

---

[1]3M Bedford Park has been selected to participate in the federal Excellence in Leadership XL program, which is part of the Clinton Administration's initiative to reinvent environmental regulations. As a participant in this program, 3M has been working with the U.S. EPA and the Illinois EPA to develop a Project XL Agreement for its Bedford Park plant. This agreement will reflect 3M's Emissions Reduction Market System reduction.

ments. The agency's proposal to allow sources to use their 1994–
1996 VOM emissions to determine their emission baseline for this
trading program is a positive step toward the attainment of these
goals. Even so, 3M believes that the baseline period for this trading
system should be more broadly defined to allow sources to use VOM
emissions from any two years between 1991–1997 to determine their
emission baseline.[2]

Allowing sources to determine their emission baselines by selecting emissions for two representative years between 1991 and 1997,
3M believes, not only will prevent strategic behavior by sources that
would otherwise increase emissions to obtain a greater allotment but
will also allow sources to benefit from early emission reductions
beyond RoP requirements made before 1996.[3] The company further
notes that using the years 1991–1997 to determine a source's
emission baseline is consistent with the agency's proposal to allow
sources to exchange emissions, on a year-to-year basis, to account for
abnormal operations during the 1994–1996 period. The paper suggests that such emissions exchanges may be limited to three years.
There is, however, no reason to limit exchanges to this extent if a
source has accurate information concerning its post-1991 emissions.

It is 3M's belief that the inclusion of this alternative baseline
determination method in the final trading program not only will
ensure that a participating source's emission baseline reflects reliable
information concerning a source's emissions but will also prevent
sources from being penalized for early emission reductions beyond
1996 emission reduction requirements.

In conclusion, 3M commends the agency for developing a market-based compliance system. The company believes that the imple-

[2]According to the agency's third draft proposal, dated April 19, 1996, a source's
emission baseline would be calculated using the average of the two seasonal
allotment periods with the highest VOM emissions during 1994–1996. The program
further provides that a source may substitute seasonal emissions on a year-for-year
basis because of nonrepresentative conditions in 1994, 1995, or 1996 but must stay
within the period 1992–1997. While 3M supports the agency's revisions to the draft
rule, which would allow sources to use pre-1994 emissions to calculate their
baselines when emissions during 1994–1996 are not representative of normal source
operations, 3M continues to believe that allowing sources to determine their
baseline by using 1991 and subsequent years' emissions information will ensure that
the sources that have reduced emissions beyond 1996 requirements before 1996 will
be able to receive credit for these early reductions.

[3]It is unclear from the discussion of the emission baseline proposal in the paper
whether the agency still intends to calculate a source's baseline from its 1996
allowable emissions. If the agency is using 1996 allowable emissions as a baseline
for some sources, 3M believes that sources that made early reductions beyond 1996
emission reduction requirements should not be penalized for those reductions.

mentation of a well-designed cap-and-allocate system will provide Chicago area businesses with the flexibility necessary to increase production, create new jobs, and make new products. Minnesota Mining and Manufacturing is confident that the agency, following consideration of the comments of 3M and other interested parties, will issue a final program that will provide Chicago area businesses not only with the flexibility to operate in a severe ozone nonattainment area such as Chicago but also with the incentive to make voluntary emission reductions.

2.3

❏

AN INSIDER'S VIEW OF THE SO$_2$ ALLOWANCE TRADING LEGISLATION

William G. Rosenberg

❏ Introduction by the Editors. *The use of tradable emission permits has been so frequently on the economist's agenda and so seldom on the political agenda that the inclusion of sulfur dioxide emission allowance trading in the Clean Air Act Amendments of 1990 is worthy of both note and inquiry. Traditional regulation as described in this book has been a standard feature of much environmental legislation in the United States. The status quo is not easily broken in the complex U.S. legislative process, so that what happened in the 1990 act is both remarkable and not yet fully explained. Hausker (1992) has thrown light on the Congressional debate over the auction market component, and there has been a Ph.D. dissertation written about the process (Kete 1992). The editors were delighted to obtain an insider's view — William Rosenberg was U.S. EPA assistant administrator for air and radiation at the time the 1990 act was passed — to add to this absorbing legislative history, whose consequences continue to unfold. Mr. Rosenberg's comments were presented to the Workshop on November 18, 1992, and were updated in 1996.*

When I came onboard at the EPA in 1989, the acid rain control debate had been going on for 10 years, and it was at a stalemate. However, two fundamental changes were under way. First, Senator George Mitchell of Maine, a downwind state to the emissions of the Midwest, had become majority leader, replacing Senator Robert Byrd of West Virginia, a state with a high-sulfur coal industry and a strong miner's union. Second, George Bush had promised, in the 1988 election, to be the Environmental President and to resolve the acid rain issue.

George Bush, as Vice President, had also been the head of a deregulatory task force that heard proposals in support of a market-based approach to environmental policy. The arguments that such an approach could be cost-effective, stimulate innovations, and be flexible were very persuasive. A lot of theoretical thinking about the advantages of a market-based approach that academics, researchers, businesses, and environmentalists were discussing at conferences began to filter into the Washington policy scene.

But policy does not follow economic theory alone, as you may have noticed. The bargaining, negotiations, and compromises were difficult, complex, and numerous. Prior bills on acid rain that were in the Congress included a proposal by Senator Mitchell to put on scrubbers or switch to low-sulfur coal. The "scrubber and switch" bill was essentially defeated by Senator Byrd, then majority leader. That was the state of political play as President Bush was inaugurated in January 1989.

The President said he was going to do something about acid rain, but he was confronted with a standoff between the Midwest region as the source of acid rain precursors and the Northeast region as the primary receiver of the emissions. We at EPA came up with a program for market-based approaches, specifically sulfur dioxide allowance trading, that had the political appeal of being a least-cost strategy. Furthermore, it implied a freedom of choice strategy that would allow utilities to choose scrubbers or other technologies, as well as cleaner fuels with less sulfur, or to purchase credits from others that overcontrolled.

What interests supported low-sulfur clean fuels? The principal areas that produced clean fuels were western region coal interests and southwestern region gas interests. So what happened? These clean-fuel areas joined with affected areas in the Northeast and Southeast to achieve a majority. That was one important foundation of support for the legislation.

What's crucial in achieving political resolution is to try to get away from ideological issues and move toward a workable negotiated settlement, where everybody wins and everybody loses a little bit. The process for developing the implementation regulations illustrates this point. When EPA convened the acid rain advisory committee, which was probably EPA's most successful advisory committee, all stakeholders were represented. We talked for five or six months. It worked out, and we had a program in record time that all parties agreed not to litigate. With all the conversations about regulatory reform, market-based approaches, and cost-effectiveness, what brought some of the environmental groups onboard was the availability of computer technology to accurately measure accountability as well as the agreement on a permanent growth cap on emissions after the year 2000. That cap meant that electricity growth no longer implied emissions growth for SO_2. David Hawkins, from the Natural Resources Defense Council and a former occupant of my job, critically supported the trading and accounting system.

Let's go back to the Congressional negotiations. Everyone thought the Administration would propose an 8 million ton reduction in SO_2 emissions. The core trade-off was a 10 million ton program (an extra two million tons for the environment) in exchange for a market-based program and the growth cap.

Without the growth cap, there could be neither an effective market-based system nor effective environmental protection. The growth cap enabled the environmentalists, particularly the Environmental Defense Fund, to go along with this reform experiment. Earlier academic discussions about the cost-effectiveness and innovation stimulation benefits of market-based approaches may have been critical in persuading some participants that the emissions growth cap would not sacrifice economic growth.

Somebody suggested earlier that the electric industry bought into the program. I disagree. The organized industry (i.e., Edison Electric Institute) never formally bought into the allowance system at the start. This was done over the institute's objection, 100%. I had a very interesting conversation with a high official from the utility industry. He came into my office after we had proposed this new approach. He was a very distinguished, nice fellow, and he listed 15 things he wanted changed. He laid them out, 1 through 15. And we listened — there were about 40 people in my office. At EPA, we have open offices and people walk in and out. No secrets.

When he was all done, I said, "I tell you what — if we adopted all of your suggested changes, would you support the acid rain trading program?" You could hear a pin drop. He said, "No." And that was the last input the utility group had. I think that this is a classic case of fundamental strategic miscalculation.

When Bill Ruckelshaus came to EPA in 1984, at President Reagan's request, he came up (after a year or two) with a 3 million ton sulfur dioxide reduction program. It was rejected by the utility industry. And so they ended up with a system designed by Bill Reilly and George Mitchell instead — a 10 million ton reduction with a growth cap.

What are the public policy lessons for the industry? If you don't come forward with a creative solution that meets public concerns but, rather, succeeds only in raising the frustration level higher, you're not going to be in much of a position to complain that government is acting unreasonably. Once you get into that kind of mode, 3 becomes 10 with a cap!

For an alternate, let's look at the newspaper publishing industry. About four years ago, there was a tremendous effort to pass laws requiring publishers to recycle newsprint. There were discussions among local, state, and federal officials about how to do this. That industry saw the handwriting on the wall, in part, I think because they're media-oriented. The *Chicago Tribune* is almost up to 100% recycling now. It's as if they said to themselves, we'll develop the technology internally so that we can more easily deink the facilities and dyes, dispose of the sludge, and get the government off our back.

I think that there is a lesson to be learned—if there's a real problem out there and the public is interested in solving it, the industries that come forward with workable solutions on a voluntary basis will have those solutions adopted. The alternative is to wait for the government to do it—then, more extreme positions are likely to come into play. The latter is a high-risk strategy.

It's my view that business has turned the corner on the attitudes of the '70s and the '80s, where often the name of the game was confrontation—"Just say no!" Now business leadership is trying to find ways to develop business plans that are consistent with public policy concerns. They're doing this in large part because they think their customers want them to do it. That's more important than the government's requiring that they do it.

Turn now to how clean air implementation is working out in the early stages. It's turning out to be very successful in a variety of ways. First of all, we had estimated that the cost was going to be about $600 per ton of sulfur dioxide emission reduction. Bill Reilly, EPA administrator, told me that, when he went to the Chicago Board of Trade to work out the details of the auction, the estimated price for initial trading would be around $285 a ton. Our best estimate in 1990 was $600 a ton, based in part on one of our econometric models, which took into account estimates of the costs of gas, coal, high-sulfur coal, and technology, all frozen at the very moment we were looking into it. These estimates were, at the time, more appropriate to a control technology program that locks into costs of the moment than to an incentive system that encourages cost-effective innovations.

Anecdotally, what's happening currently is that the high-sulfur coal business, which cannot sell to anyone other than the power plant market, is discovering increased competition from gas and low-sulfur coal and some alternative energy-efficient technologies. The pressure is to reduce the price of high-sulfur coal. So, what we're seeing is the price of high-sulfur coal coming down. Low-sulfur coal is also coming down in response to competition from other cleaner fuels and cheaper scrubbers. [The 1996 allowance auction price was only $66 per ton!]

I don't know what's going to happen on the cost-effectiveness front in the final analysis, but I can tell you that market-based approaches are very significant considerations, not only in the acid rain title but also in the reformulated gasoline program, in our regulatory negotiation on coke ovens, in the toxic area where we anticipate having interpollutant trading, and in the nitrogen oxides (NO_x) area. The pollution of the '90s is NO_x. What we do with NO_x could conceivably result in the overall cost of the Clean Air Act being even lower.

The steel industry and union, as well as the Natural Resources Defense Council, various state regulatory agencies, and EPA, negotiated a settlement of the controversy over the coke oven regulation that will reduce coke emissions by more than the statutory law requirement in exchange for the coke industry's getting a 30-day rolling average, again, a market-based approach that promises considerable savings. As a trade-off for that, the coke industry will pay for a better enforcement mechanism.

Cost modelers will have lots of business estimating the cost savings from various reform approaches. I think regulatory costs will be coming down considerably. The productivity at EPA in these new areas has also increased. In the two years since the bill was passed, 39 regulations were proposed or finalized (in contrast with the traditional 5 a year).

Which brings to mind a really big key, as far as the agency is concerned, to market-based approaches — that is the ability to do the pollutant accounting. How is pollutant accounting related to market-based approaches? The essence of enforceability is accurate data on emissions from all possible sources. Pollutant accounting with good data is also one of the keys to willingness on the environmental side of negotiations to accept an agreement. There will be a tremendous effort to do the accounting right with follow-up enforcement. Penalties for emitting more than is permitted are severe, but enforcement is required for the program to be taken seriously. Major state-of-the art improvements in continuous emissions monitoring are under way and will help greatly in this regard.

Because specific emission sources require permits which, in turn, require emissions data, we will now be able to develop adequate inventories of emissions that may not be perfect but will be better than anything we've ever had before. Our ability to measure reductions in emissions, and hence compliance, is one essential feature of acid rain control that has helped secure acceptance from public interest groups.

Let me turn to a final topic: Are allowances technically private property? Allowances can be bought, sold, or banked just like private property. Yet the 1990 Act specifically says no, allowances are not private property. The intent was to pave the way for later changes in the permissible level of emissions without expensive legal suits. I think the property right issue was very much an ideological question. In order to make this thing happen, we had to get the environmentalists' support, and the idea that people had a right to pollute was not going to fly. It's a very significant issue.

When a Wisconsin utility sold their first allowances to the Tennessee Valley Authority, they wanted to do it secretly! Unfortunately, they didn't lay the groundwork from a public relations point of view. One important response after that trade was that local environmental activists were out there saying this was a license to pollute. It had a chilling impact on the trading activity within the system as a result.

The question of failing to inform the public properly did more damage to the trading system than fraudulent accounting would have if it had occurred. I think that there are a lot of people who think pollutant trading is immoral. That the trading system will cut emissions by about half, that trading will save resources the public can spend elsewhere, and that trading will introduce flexibility and innovation into regulation have not yet been fully, or even adequately, explained. Properly communicating what good ideas can mean in practice is essential to achieving any complex reform.

REFERENCES

Hausker, Karl. 1992. "The Politics and Economics of Auction Design in the Market for Sulfur Dioxide Pollution." *Journal of Policy Analysis and Management* 11(4),fall, 553–572.

Kete, Nancy. 1992. *The Politics of Markets: The Acid Rain Control Policy in the 1990 Clean Air Act Amendments.* Ph.D. Dissertation, Baltimore, MD: The Johns Hopkins University.

2.4

❑

IMPLEMENTING AN EMISSIONS TRADING PROGRAM IN AN ECONOMICALLY REGULATED INDUSTRY: LESSONS FROM THE SO$_2$ TRADING PROGRAM

*Kenneth Rose**

INTRODUCTION

❑ Title IV, "Acid Deposition Control," of the Clean Air Act Amendments of 1990 (CAAA) established a national emission allowance trading system. The allowance trading system is a market-based form of environmental regulation designed to reduce and limit sulfur dioxide (SO$_2$) emissions. This represents a significant departure from traditional forms of environmental regulation, which previously consisted of command-and-control (CAC) mechanisms rather than market-based approaches. The argument for a market-based system is that it will result in a lower cost of compliance than a CAC mechanism for the same level of emission control.[1] Estimates of the potential cost savings that could result from the SO$_2$ trading program range from $1 billion to $3 billion per year (Portney 1990; EPRI 1993).

*Kenneth Rose is a senior institute economist at The National Regulatory Research Institute (NRRI) in Columbus, OH. The views and opinions of the author do not necessarily state or reflect the views, opinions, or policies of NRRI, the National Association of Regulatory Utility Commissioners (NARUC), or funding commissions.

[1]For an overview of trading system properties and cost saving potential, see W. David Montgomery (1972), 395–418, T. H. Tietenberg (1985), and Hahn and Hester (1989).

Other national and regional market-based environmental control programs are also being used or considered. California (the South Coast Air Quality Management District) has used an emissions offset program for volatile organic compounds for over 10 years, and this program has been expanded to other pollutants more recently (Hahn and Hester 1989; National Academy of Public Administration 1994, 24–72). Also, several states in the Northeast have proposed state and multistate trading systems (Carhart 1993), and a trading system is being considered for at least two urban areas (Chicago and Houston-Galveston). National and even global carbon dioxide trading has been discussed. Eventually, much more of an electric utility's environmental compliance could be associated with market-based environmental programs.

Since the Phase I requirements for Title IV started on January 1, 1995, and utility plans for compliance must be made several years in advance (because of the lead time of some compliance options), utilities have determined and established their Phase I compliance actions. Consequently, an examination of utility strategies, the allowance market, and regulatory actions can now be undertaken. The evidence suggests that utilities have not taken full advantage of the trading program's opportunities. Although there are several reasons for this, the most significant may be that the SO_2 allowance trading system is being applied primarily to an economically regulated electric utility industry. To realize the predicted cost savings in Phase II, utility regulators will need to make adjustments to their current practices.

BENEFITS OF A TRADING SYSTEM

The benefits of a trading program can be demonstrated with the simple diagram in figure 2.4.1 of two utilities faced with various compliance choices.[2] This illustration uses a single hypothetical marginal cost structure and different reduction requirements for two firms. In reality, of course, marginal costs are different for each firm. Utility systems have, for example, different plant configurations, plant ages, and access to low-sulfur coal. This means that compliance costs and marginal emission reduction costs will be different. For ease of exposition, however, a single curve is used here for both firms. (Using a separate marginal cost curve for each firm would not

[2]A more detailed treatment is in Tietenberg (1985, chap. 3).

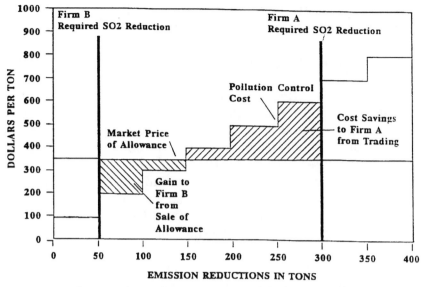

Figure 2.4.1. Selection of compliance options and the benefits of allowance trading for two hypothetical firms.

alter the results of the example.) Since the firms have different reduction requirements (because of different system configurations), they are operating at different points on the marginal emission reduction cost curve.

In this example, two utilities, firm A and firm B, have CAAA-affected units requiring a 300-ton and 50-ton SO_2 emission reduction, respectively. The environmental regulator determines for each plant an emission limit designed to meet an overall target for all emission sources. Each firm is issued an endowment of allowances based on the plant's emission limit. For the two firms shown in this example, each firm's current emissions exceed the allocation of allowances for its plants. Therefore, the firms must reduce their emissions or purchase allowances from another firm that has an excess of allowances or some combination of both. The firms can also "overcontrol" their plants and free up allowances from their allocation of allowances when they reduce their emissions beyond the required emission limit. These allowances can then be sold or saved (or banked) for future use. The figure depicts the reduction in SO_2 emissions required on the horizontal axis by the CAAA beyond each firm's respective allocation of allowances. The figure depicts the cost per ton of SO_2 reduced on the vertical axis.

Only two firms are shown in this example; however, other firms exist. All affected firms together (there will be over 2700 units affected by Phase II) determine the market price of allowances. A critical assumption is that these two firms are price takers; that is, their actions alone are insufficient to affect the market price.

Various control options, which have costs rising in step fashion, are available to the firms. In this simple example, pollution control options can reduce emissions in blocks of 50 tons with increasing incremental or marginal cost of control. To eliminate the first 50 tons of emissions requires a cost of $100 a ton with the first pollution control device. The next 50 tons of emission reductions will cost $200 a ton. The next 50 $300 a ton, and so on. The main point is that pollution control is incrementally more expensive. It can now be shown how, through allowance trading, the utility can minimize the cost of pollution control and meet the CAAA's requirements.

Firm A characterizes a buyer of allowances in this example. If the firm were to incur the entire cost of reducing its emissions by the required 300 tons, the total cost would be $105,000 (obtained by summing under the marginal cost curve: $5,000 + $10,000 + $15,000 + ⋯ + $30,000) for the first six lowest cost control options. Suppose that the market price for an emission allowance is $350. For the first 150 tons of emission reductions, the firm will choose the first three (lowest marginal cost) pollution control options for a total cost of $30,000 ($5,000 + $10,000 + $15,000). The next 150 tons, using allowances, will cost $52,500, for a total cost of $82,500. The firm saved $22,500 by reducing the first 150 tons itself and purchasing allowances for the next 150 tons. The available technology would have required an additional $75,000, but the requirement was met with an expenditure of $52,500 for allowances instead.

Firm B in the figure characterizes a seller of allowances that is required to reduce its emissions by 50 tons. In this case, the firm can meet all its required reduction with its first control option at $100 a ton for a total compliance cost of $5,000; no purchase of allowances is required. However, the next two 50-ton options can be achieved for less than the price of allowances. If the firm were to reduce its emissions by 150 tons for a total cost of $30,000, the firm would free up 100 allowances from its allocation that, if sold, would be worth $35,000 ($350 × 100). Firm B, by going beyond its required reduction, can free up these allowances in its allocation of allowances from the environmental regulator. The last 100 allowances cost the firm $25,000 to produce, for a net gain of $10,000. Since it cost the firm $5,000 to reduce the first 50 tons, the gain on the sale offsets this cost with $5,000 remaining.

It is assumed here (for the sake of clarity) that both firms have the same control costs. What varies in this example, is the required emission reduction. As noted, in reality, of course, firms have different control costs, and this, too, could cause different firm behavior, even with the same reduction requirement. Note, also, that a sufficiently high allowance price, above $700 in this example, would change firm A from a buyer to a seller of allowances.

The SO_2 trading program has already produced notable cost savings from what would have occurred with a CAC environmental program with the same level of SO_2 reduction.[3] These savings have been realized largely through *intrautility* trading (that is, utilities transferring allowances within their own system) and from the effect of competition between compliance options. A study conducted for the Electric Power Research Institute (EPRI) estimated Phase II savings attributable to intrautility trading at $1.7 billion per year (EPRI 1993, 1–18).

The following example helps explain the benefits of intrautility trading and the CAAA's flexibility. A utility builds a scrubber on one unit at a Phase I–affected plant. This results in that unit being overcontrolled and, as a result, there are excess allowances, which can be transferred to other units on the system. Under a CAC requirement, these other units may also have to be scrubbed. Now, under the CAAA allowance trading program, the utility is able to choose the unit or plant with the lowest SO_2 control cost. Savings also result from the utility's greater flexibility in choosing the option for compliance. Thus, the utility can choose among scrubbing, fuel switching, reducing utilization of affected units, repowering a unit or plant, and so on. The result is not only competition among the utility's control options but also competition between supplier's control options. For this latter reason, scrubber costs and low-sulfur coal prices have been lower than expected.

The other source of savings is trading between utilities or *interutility* trading. The Electric Power Research Institute estimated that an additional $1.2 billion in savings is available from "perfect" interutility trading (all utilities making optimal decisions). To date, as will be shown, the number of trades and the costs incurred by utilities indicate that this option, in particular, has not been exercised much by utilities.

---

[3]The magnitude of the cost savings depends, of course, on the type of CAC environmental regulation that the trading program is compared to and the amount controlled; see Tietenberg (1985, 41–47).

EMISSIONS TRADING IN PRACTICE

The above example assumes that utilities will act as a competitive firm would; that is, they will minimize their compliance cost by using the allowance market price to determine their level of emission control compared with allowance purchases or sales. However, since utilities are regulated, their actions will be influenced by the treatment that compliance costs and allowances receive from federal and state regulators. Ideally, traditional regulatory methods (which are mostly cost-based for electric utilities) may be able to induce a firm to realize the cost savings represented by the area in figure 2.4.1, "cost savings to firm A from trading." This is because, after an investment or expenditure has been made, a firm that incurs control costs above the then-current price of allowances faces the possibility of a disallowance in a prudence review. If the price cannot be anticipated, then, to avoid such a disallowance, the utility must be able to justify its inability to predict the price.

The least cost solution depends on regulators making the comparison between the market price of allowances and utility control costs. As will be discussed, with one exception, regulators have not adopted regulatory procedures that make this connection. A proper comparison of price and cost also depends on the following conditions: that there are no intentional biases (such as for in-state coal), that the planning process itself is not flawed in any way, and that the regulation does not unintentionally bias a utility's decision making. It is very likely, however, that there are flaws in the planning process and that regulation does bias utility decision making. Bohi and Burtraw (1992) point out how regulation can bias a utility's compliance decision, and Coggins and Smith (1993) discuss how the rate-making treatment can affect the allowance market.

It appears that utilities, in fact, prefer what can be described as self-sufficient strategies, preferring options (mainly switching fuels and scrubbing) that do not require purchasing allowances. Utilities, such as firm A in the above example, are, in fact, incurring much higher costs than what the allowance market could offer. Why this is occurring will be discussed after a review of utility actions and the allowance market.

In the case of firm B in the above example, the effect of regulation can clearly be seen. To comply with the CAAA, all firm B has to do is reduce emissions by 50 tons, incurring the $100 per ton cost, which is well below the price of allowances. If the revenue from selling allowances is not considered, then this appears to be the least-cost solution. However, even if the value of allowances is considered, the area in figure 2.4.1 labeled "gain to firm B from sale of allowances" would most likely, with traditional regulation, be entirely passed through to ratepayers and not shareholders. As a result, there may be little or no incentive to overcontrol and sell allowances.

Utility Phase I Compliance Decisions

Phase I compliance decisions have been made and, in many cases, approved by the appropriate state commission. By far, the preferred options are fuel switching or blending. Of 109 phase I–affected plants reported in an EPRI survey (EPRI 1993, Appendix B) out of 110 plants listed in Title IV of the CAAA, 64 are switching or blending fuel (59%). Flue-gas desulfurization (FGD or scrubbers) is next at 18 plants (almost 17%). Six plants plan to switch, or have switched, to natural gas or oil, and four are already retired, or plan to be retired, for compliance. Compliance decisions by plant are summarized in table 2.4.1.

The 21 plants that plan no action are either already in compliance (often as a result of an earlier state environmental requirement) or plan to shift allowances from a plant or plants that will overcomply within the utility's own system. Only three plants—all part of the same utility system (Illinois Power Company)—are meeting compliance primarily by acquiring allowances.[4] Other utilities are using purchased allowances to supplement what are mostly fuel-switching or blending strategies.

Actual Phase I compliance costs are difficult to come by. Often, these numbers are not reported in compliance plan filings or are presented in a manner that makes it difficult to determine whether the numbers are average or marginal cost. Estimated regional cost figures are available, however, from the EPRI study (EPRI 1993, Appendix C). The Electric Power Research Institute did not collect actual compliance cost figures for their analysis but used a simulation of individual plant and system costs for a given compliance action, combined with known Phase I compliance choices. These simulated cost figures are presented in table 2.4.2.

[4]Illinois Power Company's system compliance strategy has three main components: (1) reduced utilization of Phase I–affected units (through emissions dispatching), (2) substitution of a unit as a Phase I–affected unit, and (3) the acquisition of allowances from other sources by designating transfer and substitution units with other utilities and purchasing allowances from other sources.

Table 2.4.1. Utility Phase I compliance actions by plant

Compliance action	Number of plants
Coal switching or blending	64
No action[a]	21
Flue–gas desulfurization (FGD)[b]	18
Switch to natural gas or oil	6
Retired or retiring	4
Purchase allowances[c]	3
Total plants[d]	109[e]

Source: EPRI (1993), Appendix B, Table B-10.

[a]Compliance covered by other plant actions or already in compliance as a result of earlier action(s).
[b]Four plants are both coal-switching and building FGD facilities.
[c]Also coal blending at these plants.
[d]Plants identified in EPRI survey; other plants may also be Phase I–affected units as substitution units.
[e]Column does not sum to 109 because multiple option were being chosen at some plants.

The results show a wide variance in compliance costs across regions. Average Phase I reduction costs (1992 dollars per ton) are lowest[5] in the MAIN[6] region at $118 and highest in the NPCC[7] region at $700. Average reduction costs, however, can be misleading, particularly when such large geographic areas with many utilities are included in the calculation. Marginal costs can provide a better indication of how well the allowance program is working when compared with the allowance market price. Marginal costs (the highest regional marginal costs for selected Phase I options) are lowest in the MAPP[8] region at $332 and highest in the ECAR[9] region

[5]Excluding Texas (ERCOT, Electric Reliability Council of Texas) and western states (WSCC, Western Systems Coordinating Council), which do not have Phase I units.

[6]MAIN (Mid-American Interconnected Network) is a North American Electric Reliability Council (NERC) region consisting of Illinois, most of Wisconsin, eastern Missouri, and the upper peninsula of Michigan.

[7]NPCC (Northeast Power Coordinating Council) includes all of the New England states and most of New York State.

[8]MAPP (Mid-Continent Area Power Pool) includes Iowa, western Wisconsin, Minnesota, North Dakota, most of South Dakota, Nebraska, and extreme eastern Montana.

[9]ECAR (East Central Area Reliability Coordination Agreement) includes most of Michigan and Kentucky, Indiana, Ohio, West Virginia, and parts of western Virginia, Maryland, and Pennsylvania.

Table 2.4.2. Estimated Phase I–incurred average and marginal emission reduction costs for compliance choices in relevant NERC regions (1992 dollars per ton)

NERC Region[a]	Average Cost[b]	Marginal Cost[c]
NPCC	700	932
MAAC	439	767
ECAR	313	1,147
MAIN	118	395
MAPP	184	332
SERC	270	541
SPP	184	377
U.S.	284	1,147

Source: EPRI (1993), Appendix C, Table C-1.

[a]The North American Electric Reliability Council (NERC) includes: Northeast Power Coordinating Council (NPCC), Mid-Atlantic Area Council (MAAC), East Central Area Reliability Coordination Agreement (ECAR), Mid-American Interconnected Network, (MAIN), Mid-Continent Area Power Pool (MAPP), Southeastern Electric Reliability Council (SERC), and Southwest Power Pool (SPP). Texas (ERCOT) and western states (WSCC) are not included since they have no phase I–affected units.
[b]Includes the cost of saving or "banking" allowances for use in Phase II.
[c]Incurred marginal cost for the most expensive Phase I compliance option selected by a utility in the region.

at $1,147. This compares with an allowance price of $79 for 1995 allowances and $75 for year 2000 allowances by a recent account (Emission Exchange Corp. 1996) The ECAR regional marginal compliance cost, therefore, is more than 14 times the going rate of allowances. No region had a marginal or average reduction cost at or below the current allowance price.

These regional marginal costs can also be misleading since they reflect a single unit within the specific region that may not be representative of the entire region.[10] In a more disaggregated form, Coggins and Swinton (1994) estimated an average of the marginal costs of SO_2 control for Wisconsin utilities at almost $300 per ton[11] (Wisconsin utilities have been both buyers and sellers of allowances). These results suggest that, at the very least, some opportunities to purchase allowances have not been taken. Less aggregated data

[10]Also, these marginal costs are the incurred marginal costs for meeting Phase I compliance and do not reflect the effect of the utility redispatching its system.

[11]It should be noted that Wisconsin utilities were required to comply with state-mandated SO_2 emission reductions. This estimate reflects Wisconsin utilities' marginal cost of compliance with the state law.

(plant level, for example) would better indicate the extent to which these opportunities are being missed and higher than necessary control costs are being incurred.

The Private Allowance Market

The private allowance market (allowances purchased or sold outside the EPA auction) can be characterized as a "thin" market at this time. Most allowances traded to date have been through private transactions between buyers and sellers or with a broker or intermediary facilitating the transactions. The first allowance transactions occurred in May of 1992. Since then, the volume of allowances and number of transactions have increased slowly. As of March 1995, the total volume of private purchases (excluding the EPA auctions and intra-utility transfers) was approaching 2 million allowances of various vintages.[12] An allowance broker reports total volume at about 2.5 million allowances.[13] However, since not all transactions are reported immediately on the EPA's Allowance Tracking System, this higher number cannot be confirmed.

In either case, since there will be an average of over 5.6 million allowances each year of Phase I annual allocation to affected Phase I utilities (for 1995 through 1999) and about 9 million each year of Phase II allocation to all major emitters (for 2000 and after), the number of allowances traded so far represents a small percentage of the total allowance allocation by EPA, excluding "bonus" and auction allowances. The higher number of purchases (which most likely includes broker purchases, in effect, double-counting allowances resold to a utility) is about 3.5% of the 1995 through 2005 allocation; if the lower verified number (1.3 million) is used, bonus allowances are included,[14] and broker purchases are excluded (about 300,000 allowances), this percentage drops to 1.3%.

[12]The private allowance market activity is based on numbers supplied by Van Horn Consulting, Orinda, CA, personal communication, March 3, 1995, and Barry Solomon, U.S. Environmental Protection Agency, personal communication, March 3, 1995.

[13]Based on information from Emissions Exchange Corporation, Denver, CO, January 1995. A trend toward an increasing number of transactions is discernible.

[14]The largest category of bonus allowances is the 3.5 million allowances in the Phase I extension program for units that install a "qualifying Phase I technology" (i.e., that reduce SO_2 emissions by 90% or more, with a technology such as scrubbers) before January 1, 1997).

Of course, utilities will require many of these allowances for their operations. A U.S. General Accounting Office (GAO) study of the SO_2 allowance trading program projected the 1997 emissions of affected Phase I utilities (U.S. General Accounting Office 1994, 26), given current information on utility compliance choices, at almost 5.47 million. If this is assumed to be the approximate emission for all five years of Phase I, then the total emissions will be over 27.3 million tons of SO_2. The total five-year allocation plus Phase I extension allowances will be 31.8 million.[15] Therefore, the "excess" allowances in Phase I, that is, allowances not needed in Phase I, will be about 4.5 million. Considering only Phase I allowance purchases (various years) in the private market, and subtracting broker purchases and utility resale, there have been about 531,000 allowances purchased. This is 11.8% of the excess allowances. Therefore, thus far, about 88% of the Phase I excess allowances are being saved or banked, presumably for Phase II compliance.

The private market is characterized by several large purchasers with numerous smaller transactions. There have been nearly 70 transactions, ranging from just 12 allowances to 150,000. The leading utility purchasers are, in descending order of volume, Carolina Power & Light (CP&L), Illinois Power Company (IP), Public Service Company of Indiana, Inc. (PSI Energy), and South Carolina Electric and Gas Company (SCE&G). Combined, these four utilities account for over 61% of the total private purchases (excluding broker purchases and the resale of allowances by these utilities). These are the only utilities to purchase a total of more than 100,000 allowances each, what might be termed "compliance-sized" trades. However, only one of these utilities, Illinois Power, is using allowances for its Phase I compliance. Smaller purchases by other utilities are primarily to round out other compliance strategies, such as fuel switching. It is interesting to note that CP&L and SCE&G are utilities with only Phase II-affected plants. Nevertheless, these utilities have been acquiring both Phase I and Phase II allowances for their compliance in Phase II.

Several coal companies have also been purchasing and reselling allowances — to date, however, only in relatively small amounts (less than 20,000 total). About 36,000 allowances have been "retired" to prevent the SO_2 from being emitted into the environment (Solomon 1995).

---

[15]This is a conservative estimate of the upper legal limit in Phase I. The General Accounting Office (1994) states that the upper limit is over 7.5 million tons per year. This would be almost 38 million tons for all of Phase I.

Allowance prices have been much lower than was expected when the Clean Air Act passed, when it was widely held that they would be more than $600. As the market has developed, less price information from particular trades has been made public. From the information that has been made public, trading first began in 1992 at $250 for allowances to be used no earlier than 1995. Later trades were reported to be below $200. As noted, an allowance broker places the value of 1996 allowances at $79.[16] As will be discussed, prices dropped below $70 in the EPA auction.

EPA Auction

The EPA is required by the CAAA to hold an annual allowance auction. The allowances for the auction come from a reserve created by reducing affected sources' allocations by 2.8%. Congress created the auction provision to facilitate the development of a private allowance market. The auction is open to all interested parties, is a sealed bid auction with the sale based on the bid price, and has no minimum bid and no entry fee. Auction proceeds or unsold allowances are transferred back to affected units contributing to the reserve on a pro rata basis. Any holder of allowances may submit allowances and specify a minimum price for sale in the auction. These "private offers" are sold after the EPA-held allowances have been auctioned. The EPA is required to make public the prices and results of each auction. The CAAA specifies that EPA may terminate the auction after January 1, 2005, if less than 20% of the allowances available for purchase have been sold in any three consecutive years after 2002.

The EPA designated the Chicago Board of Trade (CBOT) to conduct the auctions. The auctions were held in March of 1993, 1994, 1995, and 1996. The auctions, by CAAA design, are split into a "spot auction" and an "advance auction." The spot auctions in 1993 through 1995 each offered 50,000 year-1995 allowances. From 1996 through 1999, 150,000 allowances from the same year will be sold in the auctions. Beginning in 2000, 100,000 allowances will be sold in the spot auction each year for the same year's allowances. The advance auction is composed of 100,000 allowances each year that are available for use in seven years (for example, in 1993, year-2000 allowances were sold; in 1994, 2001 allowances were offered, and so on). Table 2.4.3 summarizes the results of the first four EPA auctions.

[16]As reported by the Emissions Exchange Corporation, 1996.

Table 2.4.3. Summary of EPA auction results

Auction year[a]	Allowance vintage	Number of allowances sold[b]	Price range[c] ($)	Weighted average[d] ($)	IOUs[e]
1993	1995	50,000	131–450	157	8
	2000	100,010	122–310	136	4
1994	1995	50,000	150–400	159	3
	2000	25,400[f]	140–150	148	4
	2001	100,800	140–250	149	4
1995	1995	50,600	130–350	132	4
	2001	25,400[f]	128–160	131	2
	2002	100,400	126–160	128	3
1996	1996	150,000	66–300	68	2
	2002	25,000[f]	64–150	65	3
	2003	100,000	63–151	64	3

Source: Chicago Board of Trade, personal communications with author, March 1993, 1994, 1995, and 1996.
[a]Auctions occurred in March of the respective years.
[b]Includes private offers, allowances offered for sale by private holders sold after EPA-held allowances are auctioned.
[c]Price range for successful bids (rounded to nearest dollar).
[d]Weighted average for successful bids (rounded to nearest dollar).
[e]Number of investor-owned utilities that were successful bidders.
[f]Unsold allowances from the previous year's direct sale.

The EPA is required to make a portion of the allowance reserve available to nonutility power producers through a "direct sale" contingency set-aside. The price for these allowances was set by the CAAA at $1,500 per allowance in the first year (and adjusted by the consumer price index in subsequent years). As might be expected, since 1993, when these allowances were first made available, none of the 25,000 allowances in the set-aside have been sold. These unsold allowances have been sold in the following year's auction (these allowances are shown in table 2.4.3 for the years 1994 through 1996). In each auction, all the EPA-retained allowances have been sold. Only 2610 allowances from private offers have been sold in the four advance auctions, compared with 775,000 total allowances sold from the EPA reserve.

In each auction so far, relatively few investor-owned utilities (IOUs) were successful bidders (shown in the last column of table 2.4.3). The highest number of IOUs — eight — was in the 1993 spot auction. In all other auctions, there were only two to four IOUs. Of the IOUs that are successful, they often purchase a large share of the available allowances. In both the spot and advance auctions in 1993 and 1995 and the advance auction of 1996, IOUs purchased two-thirds or more of the allowances. In four of the auctions (both spot and advance in 1993, the 2001 allowances in 1995, and the 2002 allowances in 1996), 93% or more were sold to IOUs. However, in four auctions (all three in 1994 and the spot in 1996), IOU purchases ranged from just 9% to 26%.

The relatively small number of utilities involved in the auctions is surprising, given that about 40 investor-owned electric utilities are affected in Phase I and nearly all utilities (about 250 companies) are affected in Phase II. How many utilities submitted bids but were unsuccessful is not publicly disclosed.[17] However, given that the clearing price is substantially below utility costs, the few utilities involved in trading indicate that many have not been using the auction opportunity.

The price in the auctions has been lower than the price in the private market. In general, this reflects the caution used by participants when selling their own allowances and using a "bargain hunting" manner when bidding. Utilities may believe that they have "more to lose" from paying too high a price than too low a price. Moreover, since many utilities are taking a self-sufficient compliance approach, they may perceive that there is little risk in not acquiring allowances.

Additional Cost Savings Potential

The GAO study of the SO_2 allowance program's performance found that, while the environmental goals of Title IV were being met or exceeded, the full cost savings potential was not occurring because beneficial trades were not taking place (U.S. General Accounting Office 1994, 29–39). Table 2.4.4 summarizes the GAO's findings of *additional* cost savings from more interutility trading by state or region. The report found the additional national annual cost saving to be over $1.1 billion. These findings take into account the trades that had occurred by the time the analysis was conducted (mid-1994) and are consistent with the earlier EPRI findings discussed above.

---

[17]Also not publicly disclosed is the identity of a large buyer of allowances in 1994 auctions, Allowance Holdings Corporation. This bidder purchased 25.9%, 78.7%, and 89.3% of the total allowances in the 1995, 2000, and 2001 auctions, respectively.

Table 2.4.4. Estimated additional cost savings in the year 2002 from increased trading (millions of 1992 dollars)

State or region	Additional potential savings
Pennsylvania	135
Indiana	83
New York	78[a]
Florida	60
Delaware, New Jersey, Maryland, District of Columbia	58
Illinois	50
Wisconsin	50[a]
Alabama	49
North Carolina	45
Louisiana, Mississippi	45
Texas	42[a]
Kentucky	39
Ohio	39
North and South Dakota	35[a]
California, Nevada	33[a]
Arkansas	32[a]
Total	1,152[b]

Source: U.S. General Accounting Office (1994, Appendix I).

[a]These states or regions would benefit as potential sellers of allowances. The savings represent the net gain from sales of allowances.

[b]Only states with potential savings of over $30 million are shown; therefore, the column does not sum to the total shown.

IS THERE A PROBLEM WITH THE SO$_2$ TRADING SYSTEM?

In theory, it is expected that the price of allowances should reflect the marginal cost of the last tons of SO$_2$ removed. In reality, of course, perfection should not be expected since other intervening considerations will, in effect, prevent some economic trading decisions from being made. As noted, local politics has played a significant role in determining which compliance options are considered and adopted by utilities. Coal miners (in some cases, with the help of electric utilities) in many Phase I states were able to persuade legislators to pass legislation that favored the continued use of in-state coal, which is discussed below.

Of course, it could simply be accepted that there will be only one or two dozen trades a year and that the low price reflects their value to electric suppliers. As noted, in the ideal case, the market price of allowances should reflect the marginal cost of SO_2 control of all producers; or, stated differently, no utility should be incurring a higher marginal control cost for SO_2 reduction than the going price of allowances. Currently, this is clearly not the case. Again, perfection should not be expected, but the fact that allowances are selling for less than $80 while some utilities are incurring compliance costs several times this amount suggests that there is a problem.

The evidence indicates, therefore, that the full benefits of the trading program are not currently being realized. The question then is: Why, at these relatively low allowance prices (compared with incurred SO_2 reduction costs), has there been a low level of utility participation in the market to date?

REASONS FOR LOW LEVEL OF UTILITY
MARKET ACTIVITY

Several reasons have been given for the low level of allowance market activity. One possible explanation is that there is an inherent reluctance on the part of utilities to attempt such a novel and untried compliance approach as trading allowances. It is common in regulatory discussions to refer to utilities as "risk-averse" and prone to take least-risk rather than least-cost approaches to problems.[18] While this may or may not be the case, several other factors could contribute to the reluctance of utilities to trade allowances and their preference for what they perceive as a lower-risk strategy. At least five of these

[18]Utilities have argued that uncertainty in the allowance market has forced them to incur these higher control costs so that they can bank allowances for future use. However, there is no practical difference between a purchased allowance and one generated by a utility. A utility can build a reserve of allowances, as some are doing, by purchasing allowances. Building the capability internally should be done only when it is cost-effective to do so.

other factors could be causing this low level of activity in the allowance market and the utilities' reluctance to participate in it.

First, a number of states, including states with substantial Phase I compliance requirements,[19] passed legislation that was designed to encourage their utilities to continue using local coal. This was done in several ways, including preapproval provisions for scrubbers, tax credits for in-state coal, automatic passthrough of costs, and mandate of a compliance action. These actions have had the effect of limiting a utility's compliance options or biasing its decision toward self-sufficiency. Interestingly, none of these legislative actions referred to allowance purchases as an option, which would, in many cases, allow the continued use of local coal.[20]

A second possible reason for the lack of market activity was the negative press the first few trades received. News stories in *The New York Times*, *The Wall Street Journal*, Associated Press, and local papers often characterized the trades as the selling or buying of the "right to pollute." While there is an element of truth in this portrayal, it left the impression that the trades were conducted at the expense of the environment. These news stories rarely mentioned, for example, the 10 million ton annual reduction in SO_2 emissions mandated by Title IV or that the Title I requirements (the National Ambient Air Quality Standards) take precedence over the number of allowances held. Utilities, fearing negative publicity, may be reluctant to trade allowances in such an emotionally charged atmosphere. More recent trades, however, have not garnered such negative attention.

A third and often cited reason is the federal EPA rule-making uncertainty, particularly the Allowance Tracking System (ATS) and substitution rule change.[21] However, while the substitution rule change may prevent or stop some arrangements, it is difficult to conclude that it had a significant impact on the trading of allowances. Also, EPA now has the ATS operating. At worst, these were only minor deterrents to trading.

---

[19]Including, Illinois, Ohio, Pennsylvania, Indiana, and Kentucky. See Rose et al. (1993, Appendix B).

[20]One utility, Illinois Power, that was subject to a state legislative mandate discontinued construction of a scrubber and deferred until Phase II its decision to continue it. The Illinois law was subsequently overturned by *Alliance for Clean Coal v. Ellen C. Craig et al.*, case 93-C-4391 (N.D. Ill., December 15, 1993).

[21]The substitution provision of Title IV allows Phase I–affected utilities to designate other units as substitutes that then become subject to Phase I emissions requirements.

A fourth possibility is the tax treatment of allowances. The Internal Revenue Service has decided to use the historical cost basis of the allowances for tax purposes (allowances are initially allocated free of charge).[22] Some argue that this will make utilities reluctant to sell allowances since almost one-third of the revenue from the sale will be taxed. This is likely to have an impact on the sellers of allowances, who will consider the after-tax consequences of their decisions (probably requiring a higher selling price). Buyers, however, are unaffected by the IRS ruling since a tax event occurs only when allowances are sold. To date, the allowance market's problem has not been too few sellers but too few buyers. That is, as noted, utilities are forgoing economic trading opportunities and incurring marginal costs significantly above allowance prices. Utilities that have made the decision to purchase allowances have not faced insurmountable difficulties to do so. Therefore, while the IRS ruling will have an impact on selling decisions and is likely to affect the price of allowances, it is unlikely that it is a major factor contributing to the low level of market activity to date.

The fifth and possibly biggest single factor influencing the allowance market is the procedures that public utility commissions and the Federal Energy Regulatory Commission (FERC) have chosen to deal with the allowance system. As noted, regulation can distort the incentives utilities receive. Regulatory action to date has been largely reactive, responding mostly to utility compliance plans. With few exceptions, regulators have not actively encouraged utilities in their jurisdiction to factor allowances into their decision-making process.

SUMMARY OF REGULATORY ACTION

Since the CAAA was passed, state commissions and the FERC have begun to form regulatory policies or to react to specific utility actions.[23] To date, state commission actions have mostly been in reaction to utility activity. For example, the most common response by state public utility commissions has been to review and, in some

---

[22]Internal Revenue Service, Revenue Ruling 92-16 and Revenue Procedure 92-91. Tax implications are discussed in K. Rose et al. (1992, 95–101).

[23]Public utility commission and legislative activities are summarized in chapter 2 and Appendix B of Rose et al. (1993).

cases, approve utility compliance plans.[24] Utilities have submitted their compliance plans for approval or review and, after review and perhaps some modification, most commissions with Phase I utilities have responded. Often this review process has been part of an integrated-resource or least-cost planning (IRP) process. Some utilities with relatively sizable compliance costs have submitted a separate plan that considers proposed compliance actions only. Utilities regard approval or some form of acceptance of their plan by their regulators as an important step in ensuring cost recovery from ratepayers. In many cases, these compliance costs are in the hundreds of millions of dollars and represent a considerable expenditure relative to the utility's size.

Several state legislatures have enacted laws that require their public utility commission to consider — and, in some cases, approve in advance — utility compliance plans.[25] Sometimes, the commissions are restricted when, as discussed previously, the legislation is designed to encourage continued in-state coal use.

Some commissions have issued rules or orders after a utility has been involved in an allowance transaction or has requested approval to enter a transaction (for example, Connecticut, Illinois, and Wisconsin). Because allowance market activity has been limited to a few utilities, there are currently only a few responses to examine. However, these actions may be an indication of the type of actions that other state commissions will take when transactions are carried out in their states by their jurisdictional utilities.

A few state commissions (Georgia, Iowa, Ohio, and Pennsylvania) have issued general guidelines on the rate-making and/or accounting treatment of allowances and compliance costs. The Ohio and Wisconsin commissions have indicated that they are willing to consider incentive proposals from their utilities on the rate-making treatment of allowances and compliance costs (designed to encourage CAAA compliance cost minimization). Connecticut did allow a utility to retain a portion (15%) of the revenue from a sale of allowances to give the utility an incentive to seek the best price, which it may not have when all the revenue from the sale is passed through to

---

[24]The term *compliance plan* is used here to refer to a utility's plan for compliance with the CAAA that is submitted to a commission for review or approval. This may or may not be the same plan that is submitted to the federal or state EPA as required under the CAAA. The federal and state environmental regulators examine the plans for environmental compliance; the state utility commissions focus more than environmental regulators on costs that will be passed through to ratepayers.

[25]For example, Florida, Illinois, Indiana, Kentucky, Ohio, and Pennsylvania.

ratepayers. And two Indiana utilities did propose incentive mechanisms for the sale of allowances from plants that are installing
scrubbers. From the actions so far, several preliminary conclusions
can be drawn concerning rate treatment of compliance costs and
allowances. First, there is a distinct preference for automatic pass-
through of either compliance costs, allowance costs specifically,
and/or gains and losses on allowance sales. This is done either
through a fuel-adjustment type of mechanism or a compliance sur-
charge. Gains on the sale of allowances in most cases flow through
to ratepayers. Second, several states have chosen to use methods
first proposed by the FERC accounting rule as the basis for determining rate-making treatment—in particular, the use of the historical cost basis of allowances and the weighted-average inventory
method.

Third, states with utilities that sell allowances (out of state, in
particular) have been more careful in reviewing transactions than
states that have purchasing utilities.[26] Although, as noted, because of
the limited trading activity to date, there are only a limited number
of commission reactions to examine.

Finally, with respect to the review process, compliance costs
and allowance transactions have been dealt with on a case-by-
case basis. All states presently review and are likely to continue to
review compliance plans on a utility-by-utility basis and allowance
transactions on a transaction-by-transaction basis. These include
stand-alone compliance plans and plans that are part of an integrated-resource planning process. This process takes a great deal of
time and effort on the part of the commission, the utility, and other
interested parties. They are required to file detailed documentation,
present testimony, and respond to requests for further information.
For compliance plans, this process has taken over one year in many
cases. Allowance transactions (particularly sales of allowances) have
also been subject to considerable scrutiny by state commissions.
While this may be the current standard procedure for commissions
for issues of this type, it may also be particularly ill suited to a utility
operating in a competitive market. This is because it is time-consuming and inflexible and may not encourage or permit utilities to
respond to changing market conditions.

The biggest single actor involved in the economic regulation of
compliance costs and allowances is the FERC. The Federal Energy

---

[26]See, for example, the description in Rose et al. (1993) of what Connecticut and
Wisconsin, both selling states, are doing compared to Illinois and Ohio, which are
purchasing states.

Regulating Commission issued a final rule[27] in March of 1993 on the accounting treatment of allowances. The main features of the accounting treatment of allowances are a historical cost valuation of allowances, a weighted-average cost inventory method, and a new account (Account 509) for expensing allowances. The FERC also decided to use fair market value in the valuation of allowances traded between affiliates. The rule states clearly that it is intended to be "rate-neutral" and "is not intended to promote or discourage particular CAAA compliance strategies or to prescribe the rate-making treatment for allowances."

The FERC has also issued a rule on the treatment of allowances in "coordination rates," that is, rates involving power exchanges or sales between utilities that do not require the power to meet their normal load.[28] The commission will allow recovery of "incremental" costs of emission allowances in coordination rates when the rate provides recovery of other variable costs on an incremental basis. The cost to replace an allowance can be used as the basis to determine incremental cost. The FERC will permit power sellers to select a market index to calculate allowance value. The FERC policy statement specifically does not address "requirements" sales or transfers of power in pooling arrangements. The FERC states that a generic rule on these types of transactions "would be difficult, if not impossible," suggesting that this will only be done on a case-by-case basis. The FERC's determination in these transactions is likely to have a greater impact on utility actions and rates than the more limited case of coordination transactions.

CONCLUSIONS

An examination of utility compliance plans reveals that (with one notable exception) utilities have chosen self-sufficient compliance strategies; that is, they have chosen compliance options that lead to

[27]Federal Energy Regulatory Commission, "Revisions to Uniform System of Accounts to Account for Allowances under the Clean Air Act Amendments of 1990 and Regulatory-Created Assets and Liabilities and to Form Nos. 1, 1-F, 2 and 2-A," Order No. 552, 18 CFR Parts 101 and 201, issued March 31, 1993.

[28]Federal Energy Regulatory Commission, "Policy Statement and Interim Rule Regarding Ratemaking Treatment of the Cost of Emissions Allowances in Coordination Rates," 18 CFR Parts 2 and 35, issued December 15, 1994. The FERC also issued "Order Disclaiming Jurisdiction Over the Sale or Transfer of Emissions Allowances Under Sections 203 and 205 of the Federal Power Act," issued December 15, 1994. This means that the sale or transfer of allowances will not be subject to direct FERC review when the sale occurs independent of a sale of electric energy.

their own system compliance and are not utilizing the allowance market to choose compliance options or take advantage of trading opportunities. The result is that Phase I–affected utilities are incurring much higher marginal control costs than necessary (up to 14 times the current market price in one region). It is reasonable, therefore, to assume that the full benefits of the trading system are not yet being realized.

The most common state commission activity to date has been the review and, in many cases, approval of utility compliance plans. Nearly all states with Phase I–affected units have reviewed compliance plans, either as part of a broader integrated-resource plan or as a separate compliance plan. With respect to the rate-making treatment of allowances and compliance costs, those commissions that have indicated a preference have chosen to use automatic pass-through provisions in many cases. Also, in most cases (again, where the issue has been addressed), commissions have indicated that the revenue or gain from the sale of allowances should be given exclusively to ratepayers. In general, commissions are applying traditional rate-making measures (rate base/rate of return or cost-based regulatory methods) to implement the allowance program.

A traditional regulatory approach does not encourage utilities to use the allowance system in the best interess of ratepayers. For utilities that have relatively high control costs and compliance requirements, traditional regulation does not encourage a utility to minimize its compliance cost, including purchasing allowances when it is cost-effective. A utility in this situation is more likely to favor a self-sufficient compliance strategy, which presents fewer market risks and for which costs are likely to be passed through to ratepayers. For utilities that have low marginal control costs and emission reduction requirements, that is, utilities that have an opportunity to sell allowances cost-effectively, there is little incentive to incur the risk this type of strategy would entail. Such a utility may fear that it would realize little or none of the benefits and that the additional costs might not be recoverable.

Evidence that traditional regulation to date has not meshed very well with CAAA implementation includes utility Phase I compliance decisions with marginal compliance costs substantially above the market price of allowances and the fact that few utilities have taken the opportunity to purchase allowances. As noted, for the most part, utilities have chosen to generate and use allowances within their own system and are eschewing the allowance market, forgoing the opportunity to sell to or purchase allowances from outside sources.

Thus far, only one commission has adopted a review and rate-making procedure that establishes a link between the market price of allowances and compliance costs.[29] In general, the issues of finding a least-cost compliance plan and determining a rate-making treatment have been dealt with separately. There are, however, alternative regulatory procedures to traditional approaches that do make this link between costs and the allowance market and may, therefore, be more compatible with the allowance system.[30]

These methods can mimic the incentives that a nonregulated firm would receive and encourage utilities to use the allowance market more appropriately.

REFERENCES

Bohi, D. R., and D. Burtraw. 1992. "Utility Investment Behavior and the Emission Trading Market." *Resources and Energy* 14:129–153.

Carhart, Bruce S. 1992. "Emissions Offset Trading Programs in the Northeast and Mid-Atlantic States." In *Cost Effective Control of Urban Smog,* edited by R. F. Kosobud, W. A. Testa, and D. A. Hanson. Chicago: Federal Reserve Bank of Chicago, pp. 143–148.

Coggins, Jay S., and John R. Swinton. 1994. "The Price of Pollution: A Dual Approach to Valuing SO_2 Allowances." University of Wisconsin-Madison, Department of Agricultural Economics. Unpublished paper. June.

Coggins, Jay S., and Vincent H. Smith. 1993. "Some Welfare Effects of Emission Allowance Trading in a Twice-Regulated Industry." *Journal of Environmental Economics and Management* 25:275–297.

EPRI (Electric Power Research Institute). 1993. *Integrated Analysis of Fuel, Technology and Emission Allowance Markets: Electric Utility Responses to the Clean Air Act Amendments of 1990.* EPRI TR 102510. Palo Alto, CA: Electric Power Research Institute.

Emissions Exchange Corporation. 1996. "Exchange Value for SO_2 Allowances," Emission Exchange Corporation. Denver, CO. May 31.

Hahn, Robert W., and Gordon L. Hester. 1989. "Where Did All the Markets Go? An Analysis of EPA's Emissions Trading Program." *Yale Journal on Regulation* 6(1).

Montgomery, W. David. 1972. "Markets in Licenses and Efficient Pollution Control Programs." *Journal of Economic Theory* (5): 395–418.

[29]The Georgia Public Service Commission has issued an order that requires its utilities to monitor the allowance market and purchase allowances when the price is below their compliance costs or justify why they are not doing so (Order, In Re Notice of Inquiry, Review of Trading and Usage of, and the Accounting Treatment for, Emissions Allowances by Electrical Utilities in Georgia, Docket No. 4152-U (Ga PSC, April 5, 1994), p. 12). This order applies only to future compliance options.

[30]Traditional and incentive regulatory methods are discussed in detail in Rose et al. (1993).

National Academy of Public Administration. 1994. *The Environment Goes to Market: The Implementation of Economic Incentives for Pollution Control.* Washington, DC: National Academy of Public Administration.

Portney, Paul R. 1990. "Policy Watch: Economics and the Clean Air Act." *Journal of Economic Perspectives* 4(4):173–181.

Rose, K., A. Taylor, and M. Harunuzzaman. 1993. *Regulatory Treatment of Electric Utility Clean Air Act Compliance Strategies, Costs, and Emission Allowances.* Columbus, OH: The National Regulatory Research Institute.

Rose, K., Robert E. Burns, Jay S. Coggins, M. Harunuzzam, and Timothy W. Viezer. 1992. *Public Utility Commission Implementation of the Clean Air Act's Allowance Trading Program.* Columbus, OH: The National Regulatory Research Institute.

Solomon, Barry. 1995. "The Geography of SO_2 Emissions Trading." Presented at the 91st Annual Meeting of the Association of American Geographers, Chicago. Draft.

Tietenberg, T. H. 1985. *Emissions Trading: An Exercise in Reforming Pollution Policy.* Washington, DC: Resources for the Future, Inc.

U.S. General Accounting Office (GAO). 1994. *Allowance Trading Offers an Opportunity to Reduce Emissions at Less Cost.* GAO/RCED-95-30. Washington, DC.

2.41

❏

DISCUSSANT

Karl A. McDermott

THE EMERGENT EMISSIONS
TRADING MARKET

❏ While the Clean Air Act Amendments of 1990 (CAAA) created a new commodity—the "excess allowance"—that may be traded, this legislation did not create a market per se but, rather, established a minimal institutional framework to guide the EPA in creating a working market. Full-blown markets, however, do not spring forth like Pallas Athena—fully formed from Zeus's head; they generally evolve through time.

This comment will address my concerns about these new institutions from an emergent market perspective. Emergent behavior is a topic associated with the theories of chaos and complexity that are currently in vogue. It addresses the questions associated with the behavior of dynamic systems that evolve through feedback processes. Given that no formal market exists for emissions allowance trading on a comprehensive basis and that compliance under Phase I is just under way, it can be argued that current emissions trading activity is rather robust under these conditions.

My first concern is that of market imperfections. At the start, I think one point should be made clear: The Clean Air Act Amendments legislation is about abatement first and markets second. Markets are not an end; they are a means of cost-effectively achieving abatement. Many observers of the current status of emissions trading lament that trading is not proceeding fast enough and that it is not being allowed to work. This reflects a misplaced emphasis. Somebody first must undertake abatement activities in order to generate surplus allowances before meaningful trading can occur.

This brings me to a second concern about an emergent market. In order to determine whether or not you should be a seller or buyer of allowances, you must know the price of an allowance. You can know the price, however, only if there are already buyers and sellers in a marketplace. This circularity problem implies that, rather than a market instantaneously forming, price information will develop through trading processes that involve bilateral discussions and estimates of abatement costs in order to identify cost-effective abatement strategies. I employ the term *trading processes* specifically to avoid the use of the term *markets* because, in most instances, these trades are not executed in any formal market setting.

Does this imply that market imperfections are creating a problem for trading allowances? My answer to this is no, especially if you accept that, for markets to form, there must be an incentive to engage in exchange. In this case, each utility has information on each of its units' abatement costs and, if they are believed to be high, the utility has an incentive to contact neighboring utilities to explore trading opportunities. The first requests for proposals (RFPs) seeking information on the willingness of buyers and sellers to exchange allowances were issued as a direct consequence of this fact.

As a result, a number of trading processes were initiated involving bilateral exchanges ranging from cash-spot transactions to the trading of options on future allowances. Where utilities found that they possessed low-cost compliance options, they engaged in offset transactions, in effect, adopting a bubble concept for controlling multiple emission sources. Least-emission dispatching and the pooling of allowances to enable power pools to engage in least-emission dispatching are further examples of trading processes that do not involve market institutions per se.

Thus, market imperfections do not necessarily inhibit trading but, rather, provide the impetus for further market development. The steps taken by the U.S. EPA and the Chicago Board of Trade (CBOT), for example, in creating an auction market, can facilitate development of an integrated set of cash-spot, forward, futures, and options markets where exchanges can be handled by brokers or utilities themselves.

In terms of information imperfections and uncertainties, each trading process will initially involve a wide range of imperfections. The very process of estimating compliance costs bears witness to this problem. In the past several years, the expected costs of compliance have dramatically fallen from the pre-CAAA estimates. As utilities engage in various trading processes, the price information they receive will be integrated into their compliance strategy and decision framework. It is only through experience and the resulting positive or negative feedback that an incentive for utilities to rely on one form of trading or another will be created. This is the essence of a dynamic evolving market.

Regulators should be — and, I believe, are — anticipating the development of markets because they provide the type of price information necessary to evaluate utility performance. Here, again, there is a problem of circularity. Regulators would be more willing to provide incentives for utilities to use the market if they believed that the price information was accurate and that it reflected efficient trades. But the fact that many trades have been bilateral or brokered gives regulators less confidence in the efficiency of these trades. Without the ability to accurately assess performance and the benefits of trading, regulators feel uncertain about their ability to protect the public interest. Hence, regulators are reluctant to endorse markets until markets are better developed.

The process of performance evaluation also raises problems when both market and information imperfections are considered jointly. Not all utility needs could be met with standard spot or forward market transactions. The idiosyncratic nature of some utilities' needs, brought about by the configuration of capacity, fuel type, and unit vintage, gives rise to idiosyncratic transactions. These may, of necessity, occur on a bilateral basis in what could be characterized as thin markets. To the extent that competitive bidding processes can be used, regulators may feel more confident; but the general paucity of data makes performance comparisons difficult.

Thus, in a sense, emission allowances are not a homogeneous commodity, and such homogeneity constitutes one of the basic assumptions of competitive market theory. Regulators cannot, and should not, view all price and transaction data as comparable for purposes of performance evaluation. The differences in trading circumstances require that regulators develop a variety of performance standards to fit the circumstances of each transaction.

Uncertainty, as we have noted, is pervasive in the early stages of any market's development. This may help to explain the prevalence of intrautility trades and the types of compliance activities occurring in Phase I of the CAAA.[1] For example, a utility facing uncertain

---

[1]By *uncertainty*, it should be noted that I am referring to uncertainty beyond that inherent in the fact that allowances are not considered property rights under the CAAA.

market prices may find an intrautility trade less risky, or it can be argued that a utility will look at the need for allowances as an input to the production process and vertically integrate in order to supply its own allowances and avoid the classic double-margin problem. Backward integration gives the utility control over both quantity risk and transfer price (price risk).

Looking at this problem in another way, we can see that producing allowances is a form of self-insurance. When a utility needs allowances to meet its annual emissions or fluctuations in emissions due to demand or unit outages, or simply to cover its expected demand growth, the utility may be able to meet its own idiosyncratic needs more effectively than through reliance on a market transaction. If the cost/price of this option is higher than the price of allowances on the market at any time, regulators must still evaluate this decision on its total effect. Duplication of the insurance provided by self-supply may require purchasing more allowances than were used. The complexity of the choice simply provides a further example of why performance-based regulations should be adopted that place the risk and reward for these decisions with the utilities alone.

The current Phase I compliance choices seem to reflect this desire to, in part, self-supply and to keep open the utilities' future options as fuel prices and the costs of new technology change. Given the uncertainty of allowance prices, fuel switching and scrubbing in some units in order to keep other units flexible have predominated.[2] Very few utilities have relied on trading as a predominant compliance strategy.[3]

According to many observers, regulatory constraints have been the primary cause of the perceived failure of the allowance market. A regulation that may be viewed as inhibiting is one that dictates that customers, having paid for compliance, rather than shareholders, should receive the revenues from allowances. The potential for hindsight review of utility allowance transactions is also viewed as an impediment. But, to my mind, the market has been far from inhibited. As of May 12, 1994, over 115 transactions have been

---

[2]Of the Phase I units, 70 units or 64.3% adopted fuel switching or blending, while 18 units or 16.5% adopted scrubbers. The remaining 21 units or 19.2% required no modifications. Virtually 83% of all Phase I units have the flexibility to scrub, repower, or change fuel mix again to meet changing economic or technological factors.

[3]One of these is Illinois Power Company.

[4]The 2.3% of Phase II allowances is for the first five years of Phase II. I chose this cut-off point because it reflects 10 years of planning activity, which represents a significant time horizon, given the uncertainties utilities face.

posted on the U.S. EPA's Automated Transactions System. This represents 823,430 Phase I and 1,178,935 Phase II allowances. Even after subtracting the EPA auction allowances of 50,000 and 125,000, respectively, this amounts to 2.8% of Phase I and 2.3% of Phase II[4] allowances having been traded. This comes without a "fully functional" market and before trading is necessary to meet any compliance requirements. And this, of course, is a lower bound estimate, given that intrautility trades and options that don't require recording the trades until their execution in the future are not included. Nor does this estimate include the pooled allowances to meet power pool needs.

Far from being stifled, emissions trading seems to be progressing quite well and, from an emergent market perspective, it will be interesting to observe the evolution of bilateral trades, competitive biddings, and the CBOT's cash-spot and futures market. Can regulators do more to advance the process of trading? I believe the answer is yes. Regulations that allow profit-sharing and "at-risk" investments by utilities could serve as a catalyst to hasten the reactions in the marketplace that are already occurring. An example of an "at-risk" investment policy would be to provide for investments that are not rate-based and, therefore, not charged to ratepayers. The utility would be free to sell any extra allowances generated by these investments for a profit that would be given below-the-line treatment. In this way, utilities are free to match the costs and benefits of these compliance activities without concern over regulatory reviews.

Profit-sharing rules would also stimulate trading where utilities believed that the risks of losses warranted the actions. It also provides incentives, albeit weaker ones than those of an "at-risk" policy, to make marginal compliance investments. By adopting these regulatory policies, incentives would be created to rely on the market and stimulate trading activity. Far from being impediments, the imperfections and uncertainties that utilities face are the very incentives that will motivate them to adopt trading processes that will ultimately produce a fully functional emissions trading market.

2.42

❑

DISCUSSANT

Larry S. Brodsky

A CRITIC'S COMMENTS ON ACID RAIN CONTROL

❑ Don't call this sulfur dioxide allowance trading an experiment. I mean, that makes me shudder. My company has placed a lot of bets and money on the way this is working. This is the end of command-and-control (CAC), and I'm ready to cheer.

We can spend a lot of time and do a lot of studies trying to figure out what is the right allowance price and why the actual price isn't close enough to the right price, but the market is working.

We already have numerous trades that have taken place, but there really is no reason for there to be much of a market at this point. It isn't until later that deeper reductions must be made. Utilities selling electricity among themseves will have to deal with the cost of an allowance as a part of the dispatch equation.

And, when we have retail wheeling, everyone buying electricity on that retail market will have to deal with the fact that environmental costs are included in that dispatch equation. None of that has happened yet.

There has been much talk about uncertainty in this market. But there's no uncertainty about enforcement. I mean, there aren't going to be any of us trying to stretch the law on enforcement, certainly not with the civil and criminal penalties associated with compliance these days, but I think we are ethical people to begin with. That's my plug for ethics and business.

There are important uncertainties about what the Federal Energy Regulatory Commission, the IRS and, here in Illinois, the Illinois Commerce Commission might do about regulatory treatment of transactions. These decisions could make prices higher, or they might even make the allowances completely worthless.

Furthermore, there's no point in even dealing with SO_2 if we have to start controlling CO_2 in fossil fuel power plants. So there are some big uncertainties that are properly topics for later discussion.

I think something that needs to be brought up now is the gap between Phase I plans and reality. When we were all studying what we needed to do in developing our compliance plans, everyone was saying that $500 per allowance is probably about the right price to consider in your planning. Well, I guess we all, in some sense, paid good money to have studies done that estimated that $500 price. If you decided to go with technology solutions, you had a long lead time to be able to do that. You filed a plan that included the technology solution.

It doesn't matter that the market is clearing today at $70 or less. You believed $500 when you were doing your studies. You did your studies based on $500, and your studies were prudent based on the best information you had at the time. What we have now is an entirely different story. But, if you made a technology decision, it really is an irreversible decision.

There are a large number of rigidities that come into play once you employ the technology solution. You sacrifice flexibility. So some of us, Illinois Power being one, opted for the flexible solution. It was a risk at the time. We didn't have any idea what we would expect to see for allowance prices, but we knew we had to have flexibility.

Fuel switching gives us some flexibility, but there are capital costs associated with fuel switching in many instances as well as significant increased operating and maintenance costs. So we decided that flexibility had a premium associated with it and decided to enter the allowance market.

The best way for the market to stay free and unfettered is for regulators to stay out of the market. That's what a free market is. It has to be that way.

We see some regulators who are enlightened in that respect and others who have difficulty letting go of command-and-control even though that's what this new approach is supposed to accomplish. I, obviously, am an advocate of let the free market work in this instance.

Let me conclude with an observation. This country is going to realizae that we are spending a lot of money to control something that didn't need to be controlled even though we are using more cost-effective measures. What may happen is that policy will aim for 15 million tons of emissions a year during Phase II instead of the present goal of 8.9.

2.43

❏

DISCUSSANT

Christian J. Colton

THE ROLE OF MARKET FACILITATORS

❏ Trading in sulfur dioxide emissions allowances (EAs) continues to attract the attention and expertise developed in other financial and commodity markets. Entry into the new market of experienced market facilitators such as brokers, facilitators who help execute trades by introducing buyers and sellers, and market makers, market participants who take principal positions, has accelerated the development of EA trading and has enhanced trading efficiency. The result is better price discovery, market liquidity, and improved program cost-effectiveness.

In the absence of a formal, centralized exchange, market brokers provide an over-the-counter trading mechanism that satisfies price information and exchange requirements. The spread between prices bid and offered collapses in response to good price discovery, which improves market confidence and tends to increase trading activity and liquidity. An efficient exchange mechanism to transfer EAs is vital to the sustainability and effectiveness of the EA trading market. Without such a mechanism, the anticipated convergence of emissions reduction cost and EA prices may not occur and program costs will exceed anticipated levels.

Market makers improve liquidity by providing both bids and offers simultaneously. When EA prices were above $140 in 1994 and 1995, the market was flooded with offers; few utilities were aggressive buyers. Market maker bids combined with utility bids to add liquidity to an immature market. Although utility bids may have sustained the market, the added interest of market makers clearly accelerated trading and improved liquidity. To the extent that traders execute transactions through brokers, price discovery is improved.

Trading activity early on, however, was dominated by private placement transactions rather than broker transactions among regulated utilities that bought and sold within an established network of professional colleagues. Although this type of transaction represented a significant portion of EA trade volume, frequency and volume have tapered off in the past two years as confidence in the over-the-counter cash and forward markets continues to rise, liquidity improves, and external transaction costs fall. The frequency of direct placement trades between utilities diminished as the market provided better price discovery and trading became easier, partially as a result of broker and market maker participation.

Transaction costs play an important role in determining whether a trader executes transactions bilaterally or through a broker. Transaction costs include: maintenance of a database of buyers and sellers; man-hours searching for a buyer/seller; man-hours negotiating the terms of the transaction; legal fees; travel costs; additional costs arising from transactions consummated off market price; and broker's commission. Brokers establish a protocol for doing business that reduces much of this work and eliminates the duplication of effort necessary for accurate price discovery and effective transaction execution. In return for services rendered, brokers charge a commission. A typical commission to exchange 5000 EAs is about $1.25 per EA for each buyer and seller; that is, the broker would earn $12,500 on this transaction. This fee for doing business through brokers is not insignificant, and many utilities have established permanent in-house positions for EA traders, who maintain their own contacts and transact business directly for part or all of their EA purchases and sales.

As compliance and trading strategies became more sophisticated and new entrants into the market demanded creative investments, market makers and brokers introduced derivative investment vehicles to meet the needs of these trading strategies. Some of the more forward-thinking utilities also created new ways to invest in EAs and took advantage of market discrepancies.

Uncertainty about future EA prices and supply availability creates demand for derivatives, such as forward contracts, delivery of EAs and payment at a future date for a prearranged price, and options, the right to buy or sell allowances at a predetermined price within a predetermined time period at an agreed-on premium. For example, utilities with installed flue-gas desulfurization units might sell call options into a declining market to generate revenue from this otherwise dormant asset. Coal companies, concerned about their ability to market high-sulfur coal in future years, might buy EAs on a forward basis to avoid current capital expense and lock in current forward market prices. Brokers supply price and product information to the marketplace that allows participants to evaluate their trading strategies. Market makers, along with other principals, help create the liquidity necessary to execute these strategies.

Trades are generally executed for one of three reasons: to hedge a position, to speculate, or to arbitrage markets. Hedging is a technique used to avoid or reduce anticipated risk. A firm that forecasts a short EA position for the future could hedge the position by purchasing a forward contract for a portion of the position; buying a call option to cover against an extreme contango market (in which buyers will pay a premium to postpone the settling date); and leaving part of the position uncovered to be back-filled on the cash or forward market if prices improve. If the market rises, the forward position will look attractive and the firm can exercise the call options to save the difference between market price and exercise price plus premium. If the market price falls, the options expire worthless, much as premiums on an insurance policy expire. This portion of the strategy, along with the uncovered short position, is then purchased at the lower market prices.

Speculators profit from market movement. Unlike hedgers, who purchase or sell derivative products to avoid market exposure, speculators anticipate market movement and bet on the outcome. Speculation in the EA market plays an important role in market liquidity. For example, many speculators sold short into the forward EA market and back-filled these positions at lower prices as the market continued to decline. Many believe that one market maker back-filled part of a large short position through purchases in the 1996 EPA auction.

Arbitrageurs take advantage of price discrepancies characteristic of new markets. When a price disequilibrium exists, the arbitrageur simultaneously sells the higher price and buys the lower price and nets the difference for profit. With good price discovery, these simultaneously executed trades quickly force the disequilibrium from the market. In less liquid markets with limited available price information, however, price disequilibriums cannot easily be removed.

Emissions allowance traders at utilities and market maker firms have capitalized on arbitrage opportunities arising from time disequilibriums in the market. These opportunities arise from the fact that allowances identified by year have been allocated to utilities for the period 1995 through 2025, but cannot be used prior to their date.

Bid and offer prices for allowances of different dates will vary depending on discount rates and expected marginal control costs. Any discrepancies among these related prices can be advantageously arbitraged. If the transactions cannot be executed simultaneously, however, the transaction can be considered neither an arbitrage nor risk-free. If the transaction is executed over the counter, broker fees will significantly reduce profits.

Like other financial and commodity markets, the sulfur dioxide emissions allowance trading market has become increasingly sophisticated, providing new opportunities for market participants who promote and enhance active trading. The EA program, aimed at reducing sulfur dioxide emissions at the lowest cost, benefits from enhanced price discovery, competitive trading, and improved market liquidity provided, in part, by market makers and brokers. Utilities also add greatly to overall market interest and liquidity. Good price information allows utilities to develop effective strategies that meet compliance needs in the least costly manner. Market liquidity allows utilities to execute these strategies. Program costs fall when higher-cost emissions reduction is replaced with lower-cost EAs.

2.5

❑

THE LAKE MICHIGAN OZONE STUDY: FINDINGS AND IMPLICATIONS FOR EMISSIONS TRADING

Stephen L. Gerritson

SUMMARY OF FINDINGS

❑ Over the past five years, the Lake Michigan Air Directors Consortium has conducted research on the conditions that lead to the formation and transport of high levels of tropospheric smog ozone, a harmful pollutant. The chief finding has been a clarification of the influence of NO_x on ozone levels, notably a determination that reducing local NO_x emissions would result in significant increases in ozone levels throughout the Chicago and Milwaukee nonattainment areas. Based on this research, the consortium has concluded that smog ozone levels are most effectively controlled through a combination of local controls on VOC emissions coupled with controls on NO_x sources upwind of the Chicago nonattainment area. It has also been shown that ozone levels within the Chicago and Milwaukee nonattainment areas are more sensitive to changes in precursor pollutants transported from outside the region than to changes in local emissions, although controls on both sources will be necessary to achieve the air quality standard.

INTRODUCTION

Over the past decade, the Lake Michigan region has continued to experience violations of the federal air quality standard for ozone

137

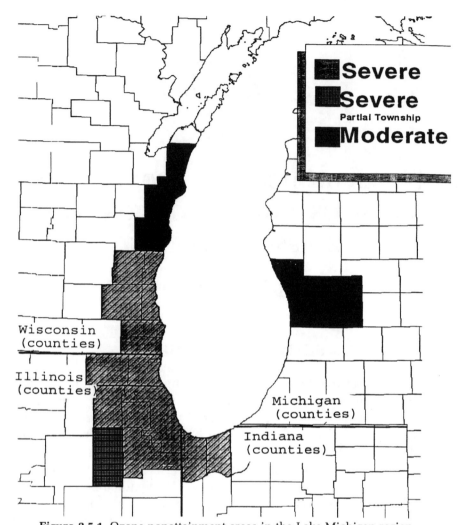

Figure 2.5.1. Ozone nonattainment areas in the Lake Michigan region.

despite the adoption of a number of federal and state control programs. Although the problem is widespread, as is shown in figure 2.5.1, it is particularly apparent in southeastern Wisconsin, where more than 10 years of compliance with federal standards and escalating levels of control have failed to reduce either the number of exceedance days or the peak ozone levels during exceedances. In 1987, there began to emerge a theory that ozone and its precursors (oxides of nitrogen, or NO_x, and volatile organic compounds, or VOCs) were being transported by the prevailing winds from the Chicago metropolitan area north to eastern Wisconsin and northeast to western Michigan. In 1989, as part of the settlement of a lawsuit[1] instituted by the state of Wisconsin to force the Environmental Protection Agency (U.S. EPA) to mandate further emission controls in northeastern Illinois and northwestern Indiana, the states of Illinois, Indiana, Michigan, and Wisconsin, and the U.S. EPA, signed a memorandum of agreement in which they set forth a plan for a regional study of ozone formation and transport in the lower Lake Michigan basin. Briefly, the primary objectives of the Lake Michigan Ozone Study were: to collect information and develop a database on emissions, meteorology, and air quality throughout the lower Lake Michigan basin; to develop and evaluate emissions, meteorological, and photochemical computer models for the prediction of ozone formation and transport; and to deliver to each of the participating states and the U.S. EPA a technically credible modeling system for predicting ozone formation and transport that could be used to test the effectiveness of various control measures in developing revised ozone control plans. In order to facilitate contracting, to create a mechanism for the receipt of grants, and to guarantee impartiality in the treatment of several controversial issues, a nonprofit research organization, the Lake Michigan Air Directors Consortium, was created in December of 1989, and the research effort began in the spring of 1990.

FIELD RESEARCH AND MODEL DEVELOPMENT

With technical and policy oversight from the states and U.S. EPA, the consortium and its contractors designed and conducted a major field research program; compiled and analyzed a database that is widely acknowledged to be among the most comprehensive of its kind; and designed, developed, and extensively evaluated computer modeling systems for emissions, meteorology, and photochemistry — models that have advanced the state of the art in these fields.

Ozone is an unstable form of oxygen which, in high concentrations, is harmful to human and animal lung tissue, and which aggravates the effects of other respiratory problems and diseases. Particularly at risk from exposure to high concentrations are asthmatics, the elderly, and young children. Studies conducted in the 1980s for the U.S. EPA on the effects of ozone exposure led to the establishment of the National Ambient Air Quality Standard for

---

[1]*Wisconsin* v. *Reilly*, U.S. Dist. Ct. (E.D. Wisc., 1989), Case 87-C-0395.

ozone of 120 parts per billion (ppb) for 1 hour. Exposure to levels higher than 120 ppb for 1 hour or more was determined to be unsafe.[2] Ozone also damages plants and causes accelerated deterioration of certain materials such as paints, rubber, and plastics.

In the real world, ozone is created in the atmosphere when NO_x and VOC molecules are broken down by ultraviolet light from the sun, and parts of these compounds join together to form O_3, the ozone molecule. Since ozone is a relatively unstable compound, an individual molecule may, in turn, be broken down by the sun's energy in as short a time as 2 or 3 minutes. At most times of the year, therefore, ozone levels are relatively low because of natural processes.

On hot summer days, however, certain conditions can lead to the creation of ozone much more quickly than it breaks down, causing the overall levels of ozone in the atmosphere to reach unacceptable levels. These conditions, which include temperatures of over 90°F, high humidity, reduced visibility or "smog," and stagnant air or light winds, result from a combination of meteorological and geographic features found in this region, usually when a high-pressure area has formed and has stabilized over the eastern United States. In these instances, ozone levels continue to increase until sunset.

The Lake Michigan Ozone Study (LMOS) field research program was designed to gather as much information about ozone episodes as possible and to create a database rich enough to be used to test the computer models being developed for accuracy. The summer of 1990 was relatively cool and, although two ozone episodes were recorded, ozone levels did not exceed 120 ppb except in isolated instances. During the summer of 1991, however, conditions were hot and dry, and there were four separate episodes, with a total of 32 exceedance days and peak values above 180 ppb. Ozone levels throughout the region were recorded using continuously operating surface monitors. A wide variety of meteorological factors were also recorded on a continuous basis, including temperature, barometric pressure, wind direction and speed, cloud cover, albedo, soil moisture, dew point, and relative and absolute humidity. In addition, many ozone-monitoring sites also had continuous NO_x monitors, and air samples were taken at these locations periodically to be tested for VOC content and species. Finally, ozone and precursor levels above ground were measured using aircraft, instrumented balloons, and lidar, a process similar to radar using laser beams to measure ozone

---

[2]Some argue that exposure to levels as low as 80 ppb for longer periods (6–8 hr) may also be dangerous. See, for example, U.S. EPA (1994).

concentrations. Readings over the lake were taken by instruments on aircraft and ships.

Every effort was made during the four ozone episodes to create a day-specific emissions inventory so that all inputs would reflect reality as closely as possible. Detailed land-use maps were prepared using satellite photography so that biogenic emissions could be more accurately projected. All major stationary sources provided actual emissions for each episode day. Area source emissions were estimated using population distributions and economic activity centers, while mobile source emissions were derived from traffic models.

The field research program was successful in that it provided data from four ozone episodes (including three different meteorological regimes), which could be used to check or validate the computer modeling system under development. The research also verified the transport phenomenon through the enhanced network of surface monitors and special tracer studies (North American Weather Consultants 1991).

During and after the field research program, a separate effort was under way to develop a new generation of computer models that could successfully replicate the complex microscale meteorology, emissions mix, and photochemistry required to produce high levels of ozone. In attempting to replicate these episodes in a computer model, many different factors must be taken into account. First, the region must be divided into small sections, or grids, so that different values for a given time may be easily assigned to different locations. Next, the model must be able to account for temporal changes. Other factors include terrain features, altitude above ground, the presence of urban centers, and so on.

The modeling system that was developed has three main components. An emissions model generates estimates of amount and species of pollutant, by source, activity, and hour of the day, for all of the more than 3000 grid squares into which the domain has been divided. For example, emissions from motor vehicles are calculated using a linked roadway traffic flow model developed by the Illinois Department of Transportation using vehicle registration and traffic surveys, which provides information on number, type, and age of vehicles on every link in the network for each hour of the day, coupled with emission rates, average speed, and trip length information for each vehicle. The output is in the form of emissions estimates from vehicles, by location and time, and the results have been verified by air sampling and testing programs. Emissions from point sources are calculated using actual flow rates reported by those sources, while area source emissions are calculated using emissions factors based on population or economic activity levels. Biogenic emissions, based on a detailed land-use map created using satellite photography, are also included. Emissions inventories are continuously updated and quality assured through sampling.

The second component is a meteorological model, which creates values for wind speed, wind direction, temperature, humidity, soil moisture, and other variables for each grid square for each hour of the day. This prognostic meteorological model was developed by the University of Colorado specifically for the Lake Michigan region and includes features such as the "lake breeze" that are unique to the area.

Output from these two components is used by the photochemical model to calculate ozone levels at the surface throughout the region, again by hour of the day. The photochemical model also contains features and improvements not found in previous versions of the Urban Airshed Model, such as the ability to vary grid size, an improvement in the treatment of vertical dispersion, and the ability to track individual plumes from particular sources. A graphic representation of this modeling system is shown in figure 2.5.2.

As noted, the output of this modeling system is a set of predicted ozone values for each grid square in the region and each hour of the day, based on a given set of emissions and meteorological inputs. Before the model was used for any simulations or control strategy development, it had to be validated through the application of a series of tests, with performance parameters established by the U.S. EPA. The first step in validating the modeling system, or demonstrating that it could be used to predict ozone levels accurately, was to compare its output for a given set of inputs to what actually occurred. As noted above, four test episodes, based on data collected during the summer of 1991, were used for model validation. While the initial results did not compare well, indications were that the problem lay in the inputs rather than the software. In particular, an analysis of air samples taken during the 1991 field research program indicated that the state emissions inventories underestimated emissions from VOC sources by as much as 1000 tons per day, or about 30% of the total inventory. Speciation of the samples further indicated that most of the "missing mass" came from motor vehicles. This finding, if true, would be consistent with research in other parts of the country, where actual measurements of emissions have been compared to estimates, with the difference being at least a factor of 2. In the case of the Lake Michigan inventory, all indications were that motor vehicle emissions estimates were approximately half of actual motor vehicle emissions.

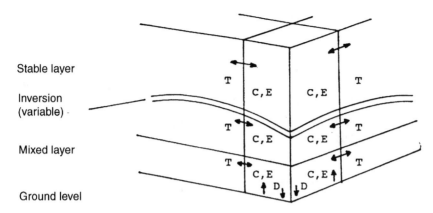

Stable layer

Inversion
(variable)

Mixed layer

Ground level

Figure 2.5.2. Conceptual representation of photochemical model. The entire modeling domain is divided into grid squares measuring 8×8 km. Each grid square receives emissions estimates (E) from the emissions model, as well as an estimate of transported ozone and its precursors (T) from all adjacent grid squares. In addition, ozone at the surface is removed through deposition (D). Meteorological inputs are provided for each grid square by the meteorological model. Finally, each grid square has a chemical reactions model (C) to calculate ozone formation based on all inputs.

Key features of the model include variable grid size and the capability to "nest" grids. For example, in urban areas, a smaller grid size is used to capture as much detail as possible from the many emissions sources while, in rural areas, a larger grid size is sufficient. A better representation of vertical diffusivity, through updrafts, radiational heating, and other factors more accurately reflects emissions inventories above the surface layer. A "plume-in-grid" feature makes it possible to track emissions from a single large source across grid squares. Depending on the meteorological conditions, the size of the vertical layers can be altered to reflect actual conditions.

In addition, biogenic emissions estimates were found to be overestimated. Emission factors for corn plants had been based on research conducted by U.S. EPA a number of years ago. Because the number of observations on which these emission factors were based was relatively low, and because even U.S. EPA was not confident of the estimates, two studies were commissioned to examine emissions from corn with some care. The results were in agreement, that emissions from corn plants were negligible. Since, according to the U.S. EPA emission factors, emissions from corn were thought to be highly reactive, this finding was quite important to the spatial distribution of ozone within the region.

Improvements in the inventory, based on more accurate measurements as well as the inclusion of previously uncounted sources, resulted in model performance that agreed quite well with observations for the test episodes, including peak levels, and temporal and spatial allocations of predicted ozone values. In short, the model was found to work quite well as a predictive tool. A series of statistical tests comparing predicted values with observations for the test episodes were conducted, and the results were submitted to U.S. EPA. The model was formally approved as a regulatory instrument in late 1994. Although approval has been granted, improvements in both the emissions inventory and in the meteorological component continue to be pursued.

SENSITIVITY TESTING

The main purpose of the model development was to provide a tool that could be used to test the relative effectiveness of a wide variety of measures controlling precursor emissions on the formation and distribution of ozone. This is done by holding meteorological inputs constant while reducing emission inputs from the particular source or sources in question by an appropriate amount. Changes in output are then compared to the baseline or test episode, and a determination of efficacy may be made.

Before actual control measures were tested, however, a series of sensitivity tests were designed and conducted to determine the response of the model to certain basic choices, such as location of the control mechanism or pollutant to be controlled. Tests were also designed to assess the relative influence of locally generated precursors and transported pollutants on ozone formation in downwind areas. It must be remembered, however that these are exercises only, to determine mode response and control direction. Reducing precursor emissions on a domainwide basis by a fixed percentage, particularly a high percentage, is politically difficult because of the stringent and direct controls required and is impossible in practice because of small-area variations.

The first sensitivity tests were designed to determine whether control of VOC only, NO_x only, or a combination of VOC and NO_x would be most effective in bringing down ozone levels throughout the nonattainment region. A series of nine tests for each control episode were conducted. (Results are summarized in Table 2.5.1.) First, VOC emissions were artificially reduced by 30%, 60%, and 90% to determine the response in levels of ozone. The results, although not unexpected, were somewhat discouraging: at 30% reductions, ozone levels domainwide decreased by less than 10 ppb. At 60%, ozone values declined by 15 to 20 ppb, a significant response but insufficient to reach the attainment standard. Only with

Table 2.5.1. Domainwide peak ozone levels resulting from changes in modeled emissions. (Results for specific locations vary with time)

Levels of reduction (%)	Precursor (ppb)		
	NO_x	VOC	Both
30	i-30[a]	d-10[b]	i-10
60	i-60	d-20	i-20
90	d-30	d-40	d-30

[a]i = increase in ozon levels.
[b]d = decrease in ozone levels.

90% reductions in VOC emissions throughout the modeling domain did ozone levels fall beneath the standard in the entire nonattainment area.

Sensitivity tests for NO_x emissions used the same reduction steps (30%, 60%, and 90%). At the 30% reduction step, ozone levels increased by as much as 30 ppb in the southern portion of the domain and showed no changes at approximately 100 km downwind (Manitowoc, Wisconsin, and northeast of Grand Rapids, Michigan). Beyond the 100-km point, ozone levels declined slightly (2–3 ppb) with a 30% reduction in NO_x emissions throughout the domain. At 60% reductions, the results were even more dramatic. Ozone levels in the lower portion of the domain increased by as much as 60 ppb, while reductions farther downwind were negligible. Only beyond the 100-km point did NO_x reductions show a beneficial impact, reducing ozone levels by 5 to 15 ppb. At 90% reduction levels, there was a predicted domainwide reduction in ozone levels below 120 ppb. Tests to distinguish between low-level and upper-level NO_x emissions showed no substantial difference in impact.[3]

As would be expected, making reductions of VOC and NO_x emissions in combination resulted in offsetting impacts: that is, the reductions in ozone levels gained from reducing VOC emissions were offset by the impacts of the NO_x reductions. Only at levels of reduction above 60% were there demonstrable benefits.

The inverse impacts of NO_x reductions on ozone levels have long been recognized. The effect of NO_x on ozone, known as "scavenging," is a result of the unstable nature of the ozone molecule and the

---

[3]Low-level or surface NO_x is generated predominantly by automobiles, while upper-level NO_x comes from large stationary sources. The chemical composition is identical.

ratio of NO_x to VOC emissions in the atmosphere. In urban areas, where the percentage of NO_x in the emissions inventory is relatively high, NO_x acts as a natural damper on ozone formation. Reducing NO_x emissions, in effect, removes the damper, allowing additional ozone to form.

While the scavenging impacts of NO_x emissions on ozone have been known for decades, they are generally limited to a very small area (10–15 km) downwind of the emission source. In the LMOS, however, these effects were seen for up to 100 km. Because of this discrepancy, it was first thought that the model's response was due to faulty chemistry codes. After independent quality assurance verified the model codes, further research on the mesoscale meteorology was commissioned. Analysis of tracer studies suggested that the "NO_x effect" was indeed real and was due to the impact of anomalies in microscale meteorology created by the presence of Lake Michigan on the nonattainment area air mass. This effect is shown graphically in figure 2.5.3.

The implications of these sensitivity tests were discouraging. First, reductions in NO_x emissions (the easier pollutant to control), would not contribute to ozone reductions in the nonattainment area and would significantly increase ozone levels in the most populated portions of the domain. Second, to meet the standard, reductions in VOC emissions of almost 90% throughout the domain would be required. Reductions of this magnitude would be impossible technologically, and even the suggestion would be politically unacceptable.

One other set of sensitivity tests was run, examining the impact of ozone levels at the upwind boundary on ozone levels within the domain. During the field studies, ozone levels at the upwind boundary were measured, both at ground level and aloft, at an average of 80 to 90 ppb and as high as 110 ppb. Because of the low level of emissions from this part of the domain, these values were used to initialize the model but were not thought to have a major impact on the atmospheric chemistry within the nonattainmnt area. Lowering the incoming ozone levels by 30% in the model, however, produced a predicted reduction in ozone levels in the nonattainment area of 10 to 15 ppb. When incoming levels of precursors were also reduced by 30%, there was an even greater response. Clearly, the entire region was being affected by transport from areas outside the domain, and indications were that a significant part of the problem in the Lake Michigan region was of external origin. Further testing led to the conclusion that NO_x reductions upwind (south of the nonattainment area), coupled with VOC reductions within the nonattainment area, would be the most efficient and effective path to attainment.

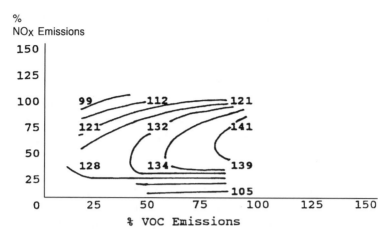

Figure 2.5.3. Impact of VOC and NO$_x$ reductions on ozone levels. This diagram shows the changes in ozone levels that result from changes in the levels of precursor emissions. (The scale is in percentages so that 100 VOC and 100 NO$_x$ represent the amount of ozone with no reductions in precursors.) The bold numbers in the diagram represent ozone levels and are in parts per billion (ppb).

As expected, reductions in VOC emissions result in reductions in ozone levels. When VOC emissions are reduced to 50% of the starting point, ozone levels drop to about 110 ppb. As NO$_x$ emissions are reduced, however, ozone levels are increased initially. It is not until NO$_x$ emissions are reduced by at least 75% that ozone levels decline below the starting point.

Although, in general, this response holds true for most of the Chicago and Milwaukee nonattainment areas, the numerical relationships differ somewhat from place to place, depending on the initial mix of precursors. This chart represents the specific response at Chiwaukee Prairie, on the Illinois-Wisconsin border at Lake Michigan.

LONG-RANGE TRANSPORT

Traditionally, efforts at reduction of ozone levels in the United States have been based on the assumption that the area showing the exceedances can correct the problem by making reductions in local emissions of precursors. While this assumption was questioned with the initiation of the Lake Michigan Ozone Study, it was completely disproved by the results of the sensitivity modeling. In response to these findings (and similar findings from other regions of the country), the U.S. EPA has acknowledged that long-range transport of ozone and its precursors was an issue that had to be addressed before the ozone standard could be attained. Accordingly, a national approach has been developed by U.S. EPA that calls for national controls on certain sources (such as commercial and consumer solvents, paints and other coatings, and small engines), coupled with a two-year cooperative research and assessment effort undertaken by U.S. EPA, all interested states, and a number of private organizations. This two-year effort, known as the Ozone Transport Assessment Group (OTAG), may result in recommendations for further controls on specific sources, which would then be implemented on a national or regional basis by U.S. EPA. Meanwhile, for purposes of preparing attainment plans, states may assume reductions in ozone and precursor levels at the upwind boundary of their modeling domains based on reasonable assumptions concerning national controls. While this policy is more realistic as to the nature of the problem, it may prove more difficult to implement, given the current political climate. Thus, the U.S. EPA and many states are also looking to nontraditional control measures, particularly those involving economic incentives to reduce emissions.

EMISSIONS TRADING

Sensitivity tests are gross indicators of directionality. In the real world, it is not feasible to reduce emissions by a fixed percentage across the board. Emissions must be reduced source by source, and control efficiencies depend on several factors, such as existing levels of control, opportunities for the application of improved technology, opportunities for changes in levels of use, costs, and so on. For example, although the most efficient and least costly means of controlling emissions from automobiles would be to decrease automobile use, in practical terms, this has proved to be quite difficult. Therefore, technological changes, such as improved catalytic converters and cleaner-burning gasoline, have been substituted at a greater cost to the consumer. As the easier improvements are made, the marginal cost of the next reduction in emissions from a given source becomes higher. Eventually, the cost curve for emissions reductions for a given source becomes so steep that further reductions are not economically feasible.

Until very recently, government agencies responsible for pollution control had assumed responsibility for assessing the potential for emissions reductions for all sources and for making determinations as to which sources should be controlled and by how much. According to industry sources, this has resulted in pollution control requirements that are extremely costly, burdensome, and inefficient. This approach, known as *command-and-control* (CAC), although flawed, has been the norm for many years.

The Clean Air Act Amendments of 1990 contained a section on economic incentives, which encourages regulators to use market forces to control emissions. Other than the SO_2 trading program and the RECLAIM program in Los Angeles, however, very little attention was paid to this section of the act until late 1994. Impending deadlines for submitting state implementation plans, coupled with a radical change in the makeup of the U.S. Congress and of state legislatures, raised questions about the traditional means of controlling emissions. Innovations, particularly involving market forces, and flexibility were encouraged.

One major result of this change in attitude has been the proposed use of emissions trading as a means of achieving the ozone standard. Rather than requiring specific sources to make specified reductions in emissions without regard for cost, a trading program sets overall requirements for emissions reductions and allows the regulated community to determine the least-cost means of achieving those reductions. Firms with high control costs would pay for reductions from sources with lower control costs. It has been estimated that the costs of reducing emissions in this manner will be less than 25% of those associated with the traditional CAC approach. The state of Illinois has established a trading program for stationary sources of VOC emissions that will be fully operational in 1999.

Economic incentive programs other than trading have also been established in the Midwest. One example is the state of Wisconsin's small engine buy-back program. The small engine (including lawnmowers, outboard engines for motorboats, and similar equipment) has never had the emission control requirements that have been placed on the automobile. In addition, because of the richer fuel/air ratio used by these engines, the amount of emissions per hour of use are orders of magnitude higher than for automobiles. Yet it is obvious that controlling emissions from such sources through traditional methods would be extremely difficult. The state of Wisconsin has instituted a voluntary program whereby owners may sell their old lawnmowers or outboard motors to the state for fair market value. Since these will be replaced with new engines, which are now subject to more stringent emissions limits placed on manufacturers, there will be a significant decrease in emissions. The costs of the program will be borne by industries for which the expenses are lower than the expenses would be for installing their own control equipment.

While the Lake Michigan Ozone Study did not create the incentives for market-based control programs, it has added a new dimension to the design of trading programs. For example, since upwind

NO_x reductions are as beneficial as local VOC reductions, it may be possible to arrange for interpollutant trades. Also, it may be possible for local NO_x sources that are required to make reductions under Title IV of the Clean Air Act Amendments to trade these requirements upwind as well.

CONCLUSIONS

The Lake Michigan Ozone Study has demonstrated the impacts of transport on the problem of ozone exceedances and has identified a potentially larger system of "superregional" transport operating over the entire eastern United States. It has documented the inverse relationship between NO_x emissions and ozone levels in the nonattainment area and has shown that controlling NO_x emissions within the Chicago metropolitan area will result in higher levels of ozone even north of Milwaukee. In addition, the model has suggested that the most effective control mechanism will be a combination of reductions in local VOC emissions, coupled with reductions of NO_x emissions upwind, to the south of the nonattainment area. Finally, a number of ways have been discussed in which economic incentive programs, including emissions trading schemes, can contribute to emissions reductions that will effectively lower ozone levels within the Lake Michigan nonattainment areas.

REFERENCES

North American Weather Consultants. 1991. "SF$_6$ Tracer Studies: Final Report." December.
U.S. Environmental Protection Agency. 1994. "Measuring Air Quality: The Pollution Standard Index." U.S. EPA Pamphlet 451/K-94-001. Washington, D.C.

2.6

❑

EMISSIONS TRADING IN NONATTAINMENT AREAS: POTENTIAL, REQUIREMENTS, AND EXISTING PROGRAMS

Daniel J. Dudek, Joseph Goffman, and Sarah M. Wade*

INTRODUCTION

❑ For over 20 years, environmental economists have been studying the potential of economic incentives to solve environmental problems. These policy tools, if appropriately designed and implemented, can deliver substantially increased efficiency in producing emissions reductions. Yet society, as represented by the general public, government, environmental groups, and business, has been slow to embrace these concepts on a widespread basis.[1] Change often involves the diffusion of new ideas, the development of human capital and institutions, and the redistribution of power and economic rents—a cumbersome and contentious process. To the extent that the use of economic incentives is viewed as "revolutionary" change, stakeholders have resisted the promise of these mechanisms.

*The authors are Daniel J. Dudek, senior economist, Environmental Defense Fund, New York, NY; Sarah Wade, senior associate, Hagler Bailly Consulting, Inc., Arlington, VA; and Joseph Goffman, senior attorney, Environmental Defense Fund, Washington, DC. The views expressed in this paper do not necessarily reflect the views of Hagler Bailly Consulting, Inc. The authors gratefully acknowledge the support of the Joyce Foundation.

[1]A recent survey of 350 corporate environmental officers found that 30% of them had no knowledge of emissions trading as an environmental compliance tool (Environmental Defense Fund 1995).

However, resistance has not meant defeat. During the last 20 years, various government bodies have experimented with economic incentives. The results have been generally positive and instructional. While these initial experiments have been very limited,[2] the last several years have witnessed a dramatic increase in the scale and number of permit markets, particularly for atmospheric pollutants.[3] The Clean Air Act Amendments of 1990, in particular, represent a dramatic departure from the existing tradition of environmental regulation in that specific market-based policies are directly authorized. Previously, legislators created programs that required regulators to specify pollution control technologies that were not necessarily cost-effective. In the face of rising concerns around the globe about economic performance, competitiveness, and cost, however, these successful policy experiments are fueling a move toward greater use of economic incentives in environmental programs throughout the world.[4]

Properly designed environmental commodity markets offer superior environmental performance and the incentives to spur innovations over time. In particular, the use of banking or overcontrol, where appropriate, creates the opportunity to produce more emissions reductions sooner than the traditional command-and-control (CAC) alternative. Banking also helps address the uncertainties of market formation and the lumpiness of control investments. Emissions trading also overcomes a traditional problem of technology-based regulation known as *grandfathering*. Since sources are given complete freedom in the choice of control strategy, there is no need for special exemptions from control for existing or aged facilities. In addition, the bias toward maintaining old, high-emitting facilities on line is eliminated. Of course, the strength of these incentives depends on the specific methodology for distributing emissions entitlements.

This paper focuses on the evolution and potential of one type of economic incentive policy: emissions trading. Marketable permits for atmospheric emissions can be seen as a natural outcome of the evolution of market-based environmental policy experiences in the United States. The transitions from lead rights trading among petro-

---

[2]For a discussion of experiences with tradable permit programs prior to the Clean Air Act Amendments of 1990, see Dudek and Tripp (1989).

[3]A path-breaking public policy study by two U.S. senators, Wirth and Heinz (1988), is generally credited with creating a bipartisan constituency for the application of market forces to environmental problems.

[4]See, for example, Dudek et al. (1993, 58–75).

leum refiners to emissions trading in criteria pollutant nonattainment areas to national SO_2 trading to transferable production permits for chlorofluorocarbons (CFCs) are very clear. This experience can help lay the foundation for international trading in greenhouse gas reductions for the control of global warming, potentially the largest of such markets. Tracing the development of these strategies through to the present can illustrate how widespread emissions trading is becoming and what future directions it may take.

Brief Description of Emissions Trading

The last few years have witnessed a phenomenal increase in the scale and number of environmental programs that are based on marketable permits and other economic incentives. Perhaps the most impressive of these "new" programs is found in the acid rain requirements (Title IV) of the 1990 U.S. Clean Air Act Amendments, which represent a dramatic departure from the existing tradition of environmental regulation. By and large, environmental programs in the United States have specified pollution control technologies. Depending on the political strength of the industry, these prescriptions might even be accompanied by subsidies. Title IV's acid rain program, in contrast, utilizes a market-based strategy to achieve SO_2 reductions.

Under the acid rain program, each polluter (initially, all sources in this program are large electric utilities) is given an endowment of emission allowances for each plant, much like initial deposits in a checking account. These allowances are freely exchangeable throughout the United States. In fact, anyone can hold these transferable permits. The allowance allocations simultaneously prescribe mandatory emission limits and the reduction responsibilities required at each location. Through the artifice of the transferability of these permits, the U.S. Congress created a valuable asset, one that is utilized as part of the financing strategy to produce compliance with the acid rain program.

Once the market realized that trading would not only reduce compliance costs but also create new revenue flows for some polluters, it responded with alacrity, giving rise to an array of ancillary institutions designed to support the SO_2 emissions market.[5] For example, there are several public and private auctions of allowances,

---

[5] For an empirical analysis of both cost savings and cost sharing associated with SO_2 allowance trading, see Dudek (1989).

with no restrictions on who may participate.[6] The Chicago Board of Trade (CBOT) has proposed a futures market for allowances. Several firms are developing insurance services based on allowance pooling. In addition, electronic bulletin boards and computerized tracking systems have been developed to facilitate trading.

While these ancillary institutions have seemingly developed overnight, it is important to note that markets for environmental discharges have been under development for well over a decade. Progress in the development of practical, functioning, supporting policies for these markets has been slow—in part because the status quo is a vicious competitor and innovations are subjected to greater scrutiny.[7] The 1990 Clean Air Act Amendments have provided the impetus to overcome that inertia.

The New Paradigm in Environmental Protection

There is a new emphasis on environmental performance and efficiency. The acid rain program was preceded by over 10 years of efforts to establish emissions trading programs using a variety of approaches. Among them, the acid rain program is perhaps the most highly developed, offering a template for current efforts to expand the application of emissions trading. Because the Clean Air Act Amendments of 1990 impose a new round of ambitious environmental obligations, Congress also provided for increased opportunities for cost savings and greater rewards for innovation through the use of economic incentives. While this strategy is not new to the problem of reducing air emissions, the emphasis and encouragement constitute a hallmark of the 1990 amendments in particular.

Earlier efforts to establish economic incentive strategies have taken many forms in air quality management. These have ranged from the relatively common tools of offsets, netting, and banking, to the emissions cap-and-trading model established in the acid rain program. The former group, under which emission reduction credits (ERCs) are created, contains the most limited form of economic incentive, with emphasis on internal rather than collective optimization. Individual firms have the option of reducing a portion of their pollution control costs and, in exchange, provide "society" with

[6]Other than the normal financial bona fides, potential participants in the public auction of SO_2 allowances conducted by the Chicago Board of Trade for the EPA need only establish an account in the allowance tracking system.

[7]For an examination of the obstacles to implementing functioning environmental commodity markets, see Dudek and Palmisano (1988).

collateral environmental benefits, such as improved inventory information, easier compliance, and accommodation of economic growth along with continuing progress in reducing emissions.

The regulatory constraints inherent in attempts to piggyback these tools onto existing programs limit the incentive for the polluter to choose them. Although optional flexibility is a feature of offsets, netting, and banking, it is not the same type of flexibility found in comprehensive emissions trading programs. Optional flexibility can actually be more difficult and, therefore, more costly for firms.

The emissions cap-and-trade model of the acid rain program is a comprehensive system that offers broad incentives to the firm and, potentially, to regulators. Under this model, emission trading drives a mandatory compliance program aimed directly at emissions reduction performance, or collective optimization. All firms receive direct emissions responsibilities backed by rigorous monitoring and stringent enforcement in exchange for complete flexibility in method of compliance. Regulators maintain a strong enforcement position and reduce some of the administrative work required to support CAC and piggyback systems. This approach puts the cost-saving imperative, reward for innovation, and superior environmental performance in a mutually reinforcing alignment. No special review process is triggered by firms wishing either to overcontrol or to transact reductions as required by the review of the generation of ERCs.

The focus on environmental performance and results rather than specific control technologies is critical to the success of this new coupling of flexibility and responsibility in environmental management. This approach strives to replace the current system, one that promotes specific technologies without ensuring commensurate emissions reductions, with a comprehensive framework in which all emission sources are subject to the same degree of strict accountability.

The problems with the current system are apparent in the current regulation of ozone precursors. Stationary sources of emissions are typically regulated with stringent emission rates or technology requirements. However, this regulation often neglects the utilization of the facility or the performance of the technology. Therefore, if the source operates more than was expected or if the technology underperforms, emissions increase while the source continues to remain in compliance with the requirement. Mobile sources have been managed through a combination of tailpipe emission limits and fuel regulations. This control strategy, which affects only the rate of emissions, cannot limit total emissions discharges from a vehicle since the number of miles driven is not controlled. The number and type of trips, as well as driving behavior, further differentiate the relative contribution to total emissions of vehicle miles traveled (VMT). Further, even the method of certification for vehicle conformity to federal tailpipe standards has come to be understood as a poor predictor of actual in-use emissions.[8]

In addition, the present system for achieving and maintaining the National Ambient Air Quality Standards (NAAQS), which include ozone, is largely dependent on states for implementation. Unfortunately, state boundaries do not follow the geography of airflows distributing pollutants. For some states, these flows are so consequential that attainment cannot be achieved without reduction of pollutant emissions outside the state. Interstate cooperation in emissions reduction has been left to the coordination of state implementation plans and U.S. EPA review.

The result of these regulatory policies is a system that cannot control total discharges and that imposes different reduction responsibilities (with different control costs and levels of accountability) on sources, sometimes depending on the political leverage or vulnerability of these sources. Of course, the environmental underperformance of technology-based strategies, even in cases in which compliance with the technology standards is achieved, is often matched by their economic underperformance. Technology- and standard-setting regulatory processes may aim to account for costs, but they fail to provide the sources themselves with the optimal latitude to rationalize their compliance costs. Because emissions trading offers polluters the cost savings provided by a full range of options for meeting their emissions reduction obligations, the compliance strategies ultimately selected are much more likely to be lower in cost through the capture of dynamic efficiencies. This flexibility accorded to sources weakens the industry's ability to dispute the technical and economic feasibility of achieving the reductions sought. Finally, given the array of environmental challenges facing industry, such cost saving becomes ever more critical, as does the assurance that the costs incurred will deliver the environmental results.

---

[8]Vehicle maintenance, driver behavior, and power enrichment are all significant determinants of actual in-use emissions from any individual vehicle. For example, hard acceleration frequently leads to commanded enrichment, i.e., the addition of extra fuel to reduce operating temperatures in order to protect the expensive catalysts at the heart of the emissions control system. The result is a significant increase in the discharge of hydrocarbons—critical ozone precursors. For a more detailed discussion of the problems in managing vehicle emissions, see Dudek (1993, 90–110).

It cannot be emphasized enough that emissions trading is not just about cost saving or more efficient regulation. Rather, it can provide a fundamental transformation in the management of environmental problems. In addition to defining compliance in terms of explicit accountability for actual emissions, the promise of emissions trading is not the immediate and obvious efficiency gains from rearranging control effort in response to differential control costs but, rather, the promise of improved environmental outcomes through new control strategies, techniques, and technologies developed over time in response to market opportunities. In addition, properly designed environmental commodity markets offer superior short-term environmental performance as well. In particular, the use of banking or overcontrol, where appropriate, creates the opportunity to produce more emissions reductions sooner than the traditional CAC alternative. This, in turn, yields a longer-term benefit as well since many overcontrol strategies involve early investment in durable emissions reduction approaches and techniques. As is the case with acid rain, these early reductions may also yield higher marginal benefits. By increasing the penetration of aggressive or innovative strategies, early investment can accelerate cost reductions and encourage the wider use of substantial reduction strategies by sources.

Emissions trading also offers the opportunity to create programs designed to lead to comprehensive emissions reduction strategies. For example, the Environmental Defense Fund (EDF), in collaboration with the General Motors Corporation, designed a vehicle-by-vehicle emissions measurement-based regulatory protocol for identifying and retiring high-emitting vehicles for their avoided emissions value (Dudek et al. 1992, U.S. Environmental Protection Agency 1993). This policy was explicitly intended to help break down the arbitrary distinction between mobile and stationary source emissions embodied in traditional nonattainment management. While a number of stationary source trading programs have been developed in a number of states, it is unlikely that such programs can produce attainment in the absence of effective mobile source emission control strategies. Other policies developed under the Economic Incentive Program (EIP) provisions of the CAAA of 1990 that address mobile and stationary integration include fleet fuel conversions and urban buses. However, these are unlikely to provide for the comprehensive integration of stationary and mobile sources necessary to achieve attainment. Lack of such integration could also impede stationary source market development alone.

Finally, emissions reduction markets increase the likelihood that economic resources will be directed toward the most fruitful innovations. Technology-based standards, in contrast, often arrest or misdirect investments.[9] Several states have made significant progress in developing emissions trading programs; others can learn from their experience. A number of these programs are reviewed in a later section of this paper.

COMPARATIVE REVIEW OF BASIC EMISSIONS TRADING MODELS

Introduction

Just as there is no universal form of CAC regulation, there is no universal model for emissions trading. Generally, all emissions trading models contain the following components: a method for assigning property rights,[10] a mechanism for exchanging these rights, and an opportunity for regulatory review and enforcement.

Each of these features can be designed to address numerous concerns, perceptions, constraints, and ideologies to form politically viable coalitions for the specific program. Consequently, the basic trading model has evolved into at least three discernible forms. To the novice and, in fact, to the experienced practitioner, these models must start to look alike, for it is often easier to place them on a continuum than it is to confine them to strict boundaries. However, such definition is important to understanding both the evolution of emissions trading and its future. Consider the following simplified models.

Emission Reduction Credit (ERC) Model

A permitted source makes a permanent emissions reduction, obtains certification of the reduction from the local air quality regulator in the form of a permit modification, trades the credit, and obtains approval for the trade from the regulator. In this system, reductions are optionally undertaken by sources and certified prior

---

[9]New source performance standards, for example, created biases against the use of cleaner fuel and energy conservation as pollution control strategies.

[10]In the context of emissions trading, property rights are defined as limited authorizations to emit that can be altered by the authorizing agency as the underlying environmental circumstances change; for example, as new science developed concerning stratospheric ozone depletion, the phaseout of ozone-depleting substances was accelerated.

to exchange. The buyer knows that the reductions purchased are quantifiable, surplus, permanent, and enforceable. Approval of the trade involves ensuring compatibility with air quality objectives. Mandatory nonattainment area offset requirements for new sources have been a primary source of demand for ERCs.

Cap-and-Trade, or Allowance, Model

A source receives an emissions allowance and is free to bank or trade this allowance but, at the end of the accounting period, it must have sufficient allowances to cover its emissions. Since the overall system includes a fixed annual limitation on the number of allowable emissions allocated to sources, specific reduction strategies do not need regulatory approval. Sources transacting in the market know immediately that allowances are fungible and fully accepted for compliance purposes without additional regulatory approvals. Each source is required to possess only allowances sufficient to cover emissions during the compliance period. Sources that do not meet this requirement pay hefty financial penalties, including compensating reductions to pay back the environmental account, a feature unique to this policy approach.

Discrete Emissions Reduction (DER) Model

A source measures and documents actual emissions reductions and trades them. The source acquiring the DERs asks the regulator for permission to use the DERs to comply with a specific reduction obligation. This system, like that for ERCs, is based on voluntary reductions by sources. However, reductions are not certified by air quality regulators at the time of creation. Rather, buyers of DERs evaluate the relative cost and quality of reductions, make a purchase, and then attempt to seek regulatory approval for the use of those specific DERs to satisfy compliance obligations.

The obvious question is: Given the common heritage of the models, why is each model so different? The answer can be found to some extent, in an analysis of the policy context that shaped these models.

The 1986 Emissions Trading Policy Statement and the ERC Model

The earliest development of emissions trading occurred in urban areas with severe air pollution problems. The Clean Air Act Amendments of 1977 set very stringent deadlines for cities to comply with ambient air quality standards for criteria pollutants. Failure of cities to meet the deadlines would have invoked federal penalties that included the loss of federal funds earmarked for the community. In order to finesse this difficult political problem, a collection of policy experiments, collectively known as emissions trading, began. The individual strategies were offsetting, bubbling, and netting. These strategies were developed as opportunities to introduce incremental flexibility into the command-and-control system in order to encourage stationary sources to reduce emissions.

Ultimately, these various attempts to establish bubbles, offsets, and netting were codified by the EPA in the 1986 Emission Trading Policy Statement (ETPS). On paper, development of this policy statement marked a positive step forward in the evolution of emissions trading in synthesizing both the experiences and legal challenges of preceding years. In reality, the story was quite different. The ETPS almost killed emissions trading.

The ETPS was fundamentally flawed in that it attempted only to add some flexibility to existing CAC regulations rather than changing the basic approach to pollution control. So, while the concept of emissions trading was hailed as revolutionary, actual trading was limited in scope, overly costly, and somewhat frowned upon by regulators and the general public. Consequently, the number of trades sanctioned by official state emissions trading programs and developed in accordance with the ETPS is relatively small.

For emission reductions to qualify as ERCs under the ETPS, they must be permanent (e.g., an equipment curtailment or installation or a process change), quantifiable (i.e., the source must have an official emission baseline and use sanctioned emission measuring methods), surplus to the reductions required by the source and/or by the state in order to attain the National Ambient Air Quality Standards (NAAQS), and enforceable (i.e., reflected in a source's state and federally enforceable permit).

When evaluating past ERC programs, it is important to consider their limitations. Under CAC programs, sources received a permit allowing them to emit at a certain rate. Typically, these permits did not establish limits on the number of hours of operation, consequently they did not establish a firm emissions limitation or cap. Many sources found that their permitted levels could be as much as 40% higher than their typical actual emissions. When a source decided to create or use an ERC, it lost all the flexibility it had in its permit with respect to the difference between allowable emissions and actual emissions. The source had to establish an actual emissions baseline and make permanent reductions from that baseline. Therefore, in order to maintain some operational flexibility, the source would not generate ERCs until it had a buyer for them, unless other reduction requirements provided the opportunity for additional low marginal cost reductions. Nor would the source have the incentive to sell ERCs that it might need in the future. The resulting market was inefficient but still generated significant cost savings.

The level of constraints and supervision inherent in the ERC model, as set forth in the ETPS, reflects an intense degree of oversight as if expecting sources to cheat the system and states to be lax in enforcing emission reduction requirements. However, in the context of the circumstances under which the ETPS was crafted, such conservatism was warranted. Public skepticism, legal assault, and lack of political support were the dominant circumstances. Under the ERC model, sources have to provide regulators with significantly more information than they do to receive a permit, and states have to provide EPA with more opportunity to review credits and trades than is typically the case with other permits. This intensive oversight has increased transaction costs. Yet, at the same time, the overall credibility of emissions trading as an environmental management tool has been significantly improved under the regime.

Cap-and-Trade, or Allowance, Model

Emission reduction credit models failed to provide sources with real incentives to participate in the trading program. They also failed to give the public a performance-based environmental management system. The requirements for firm baselines and permanent reductions decreased flexibility for sources. Also, the role and timing of government review interfered with the free market forces supposedly unleashed by marketable permits. Cap-and-trade models were designed to address these problems. The acid rain trading program is the best-known example of this model.

The basic structure of the acid rain program is quite simple. The program includes an emission cap on all electric utility sources of sulfur dioxide (SO_2), which is the total amount of SO_2 allocated each year. A portion of the total emissions cap is allocated to each source in the program. Sources are relatively free to trade allowances. At the end of each year, sources submit records to EPA proving that they did not emit more SO_2 emissions than they held in allowances. Allowances held in excess of emissions are "banked" or carried forward. Any excess emissions are penalized by an automatic penalty of $2,000 per ton, plus the requirement to produce a compensating reduction in the next year.

There are several reasons why this specific program works so well. However, these very same reasons may make it challenging to transfer the concept to other pollutants or groups of sources. First, only large utilities participate in the program. This means that the sources are relatively homogenous and easy to oversee. All the SO_2 emissions in the program come from smokestacks rather than diffuse, hard-to-monitor sources. Finally, the program is based on a critical decision to treat acid rain as a total loading problem rather than an episodic problem. This means that allowances are completely fungible in temporal and geographic terms. Since overall reductions are on the order of 50% over the life of the program, it is assumed that this reduction in total loading will improve the environment in the long run.[11] Most other pollutants are not currently treated on a total loading basis.

One of the major obstacles to implementing the acid rain program was determining the initial allocation of SO_2 allowances. In fact, many other regulatory agencies that have attempted to start cap-and-trade programs have found the allocation issue to be one of the main problems, though not an insurmountable one. Utilities do not have constant emissions over time. Plant outages, emergency situations, and weather all play a key role in determining the percentage of time that a power plant operates. In the case of the acid rain program, it was difficult to find one year that was representative for all utilities in the program. Consequently, the acid rain program allocated SO_2 allowances based on an average of historic actual emissions. An additional consideration had to do with the installation of pollution control equipment. Some sources that had earlier complied with reduction requirements felt that they deserved an amount of allowances that reflected their precontrol emissions. Otherwise, they argued, they were being penalized for complying early while their counterparts, who had not installed equipment, were benefiting from failing to comply with the law. Similar issues arose in the Regional Clean Air Incentives Market (RECLAIM) program in Los Angeles.

The allocation issue has been central to the debate about cap-and-trade systems because the allocation of emissions reduction responsibilities is overt and public rather than covert and discretionary as

---

[11]Relative cost gradients, the spatial distribution of existing emissions, prevailing winds, and the distributions of reduction responsibilities favor this outcome. For an economic evaluation of the expected pattern of net trading, see Dudek (1989).

in traditional CAC systems. In fact, public allocation of allowances has provided an opportunity to reward early adopters. Irrespective of whether firms are explicitly given allowances, the allocation of emissions reductions responsibilities is undertaken in the development and implementation of nonattainment plans. The issue is whether allocation is done explicitly in the auditorium or implicitly in the back room.

Discrete Emissions Reduction Model

The recent proposal for open-market trading is the latest evolution of the emissions trading model. This proposal, now under consideration by EPA, combines elements from the ERC and allowance models. Under the open-market model, sources generate discrete emissions reductions (DERs) by reducing emissions below permitted levels. As in the ERC model, participation is voluntary. The similarities stop there, however.

Once a source generates DERs, it is free to trade them to other sources. In this respect, DERs are treated more like control technologies than emission credits. A source wishing to purchase DERs is responsible for performing the same due diligence it would perform in deciding what type of scrubber to purchase. The sources generating DERs are responsible for documenting the reductions but, unlike ERCs, no prior certification of the reductions is performed by air quality regulators. If the source selling DERs does not have adequate documentation, then the DERs are likely to be substantially discounted. Similarly, DERs from a source that has impeccable documentation of the reduction will be more valuable. In either event, the uncertainty of acceptance by regulators will reduce their value.

Once trades are effected, regulatory agencies are required to review the trades. In reality, the first trades accomplished under the DER model may have significant regulatory review for all stages of the trade. However, once the system is established, it is expected that DERs from a pool of reductions will have essentially blanket authorization. It is also expected that, once a type of trade is approved, other similar trades can be approved more easily. Nonetheless, DERs are unlikely to be a mainstay of corporate environmental compliance because of the substantial uncertainty surrounding their specific use for compliance. If the proposed guideline on open-market trading survives in state-adopted programs, DERs are more likely to allow firms to practice emissions trading without the obligations or benefits of assuming a baseline and cap as in allowance trading.

It is hoped that the DER model will be another tool in the emission trading toolkit, but it is not clear that this will materialize. Sources can generate DERs by capturing the difference between allowable and actual emissions, a common problem and source of litigation in the early development of ERC markets. Presumably, this capture would be accompanied by self-certifying documentation describing the magnitude of reductions and methods used to obtain them. However, emission trading markets have been primarily hampered in their development not by inadequate supplies of reductions but by uncertainty on the part of buyers that acquired reductions will provide completely equivalent compliance. Caveat emptor is not a recipe for overcoming this problem. DERs, as currently proposed, would share with SO_2 allowances unlimited life, once created, without the environmental or financial benefit of a cap on total emissions since none is required by sources producing DERs. Since DERs are conceived for use in nonattainment programs, it is difficult to see how past reductions will relate to a contemporaneous environmental problem such as seasonal ozone nonattainment. In short, this proposed program offers significant flexibility but at the expense of breaking the link between reductions and solution of the underlying environmental problem.

APPLICATION OF THE BASIC MODELS: EXPERIENCE ON THE FRONTLINES

States and organizations representing groups of states developed trading programs all over the country. These trading programs were influenced by the local political culture, and by the perceived role of federal regulation and guidance. Many programs share characteristics of one of the models described above. Some draw on several models.

A number of states adopted trading programs or attempted to establish ERC markets. Most of these policy developments are incremental extensions of current experience. The following cases highlight unique policy options or obstacles faced by those states attempting to develop trading programs. This list is not an exhaustive review of current trading efforts. In some cases, genuinely new initiatives are under way that may deepen the spread of markets for emissions reductions.

Texas

In June of 1992, Texas initiated a feasibility study for a VOC and NO_x permit allowance trading program, using the eight-county Houston-Galveston ozone nonattainment area as a case study. That study has subsequently been completed. The report calls for the development of three levels of market-based incentive programs to reduce ozone pollution. The first level is concerned with the acquistion of offsets for new sources. As part of this effort, the Texas Air Control Board (TACB), currently reorganized into the Texas Natural Resources Commission, passed ERC banking rules and created area emissions reduction credit organizations (AERCOs). Level II programs will be a hybrid of command-and-control and straightforward emissions trading and will include rules for NO_x reasonably available control technology (RACT) trading, vehicle scrappage, and alternative fuels programs. Level III adopts an emission cap and allocates emission credits to existing sources. To date, Texas has passed the banking and AERCO rules. It is developing the rules necessary to implement level II programs and is waiting for level III programs.

Highlights of the banking and AERCO (otherwise called community bank) rules include the following:

- Creditable reductions must be a minimum of 10 tons per year and are prioritized in the following order: shutdowns, emissions from facilities with at least two years of continuous emissions monitoring, emissions from facilities whose permit was reviewed within the immediately preceding two years, emissions from facilities with standardized quantification procedures, and other emissions.
- AERCOs are nontaxing and nonregulating agencies that may receive funds from virtually any source. Their purpose will be to identify and promote projects and strategies to generate emission credits. Once each bank (AERCO) has a minimum of 1000 tons per year of registered ERCs, excess credits go into a waiting line to be registered when space opens up, i.e., when the credits first deposited are utilized.
- The rule explicitly allows for the creation of mobile source credits through the use of alternative fuels and scrappage programs.
- ERCs can be used only within the same designated nonattainment area.
- The ERCs will be discounted 3% per year to help with the reasonable further progress (RFP) demonstration.

The Houston area was chosen as the case study for development of this program, which will be implemented throughout the rest of the state if it is successful in Houston. As the TACB writes, it wants the move toward emissions trading to be "evolutionary, not revolutionary." The TACB is actively pursuing mobile source credit quantification procedures and has launched a 1-year, 50-vehicle pilot program to determine emission reductions through conversion to natural gas from gasoline.

Los Angeles, Regional Clean Air Incentives Market (RECLAIM)

The RECLAIM project started as the most ambitious and comprehensive nonattainment trading program in the country. The Los Angeles basin is the only extreme nonattainment area in the country. As such, it is unique in its problems and approach. Its pollution problems are compounded by population density and meteorological conditions. The South Coast Air Quality Management District (SCAQMD) embarked on RECLAIM as an alternative to its traditional CAC rule-making process. RECLAIM was to be less costly and more effective at reducing pollution. During a two-year period, RECLAIM participants developed climatic, photochemical, and economic modeling of the effects of a trading program. The effort resulted in a series of five working papers and a draft program. The program went through a public comment period and was changed dramatically prior to adoption.

The basic concept embodied in RECLAIM is that each source within the basin is allotted an emissions allocation consistent with an overall regional cap. Today, that cap exceeds the total emissions allowable when the area is in compliance with the NAAQS. Therefore, both the overall cap and each source's allocation will decline in future years until the total emissions cap for the area equals the amount allowed in accordance with the NAAQS. Each source is responsible for reducing a certain amount of pollution, through either outright reductions or the acquisition of emissions reduction credits from other sources.

Far from realizing its initial ambitions, RECLAIM today applies only to stationary sources of NO_x and SO_x that emit at least 4 tons per year—535 sources, according to RECLAIM staff estimates. Baselines will be the highest emissions during the period 1989 to 1991 reduced by an amount equal to the reductions required by the existing rules from that peak year to the start of the program. The annual emissions reduction targets are based on a straight-line rate of decline between the baseline and end point. A market for volatile organic compounds (VOC) is planned for the near future. Credits known as RTCs (RECLAIM trading credits) can be bought or sold on a yearly basis, and such transactions are reflected in permit modifications. The basin is divided into two sensitivity zones—coastal and inland. Trades are allowed within the two zones, and the coastal zone may sell to the inland zone but not vice versa. Sources must submit monthly reports and certified quarterly reports. The program began on January 1, 1994. It is estimated that RECLAIM will save Los Angeles $164.1 million, or 47% of the $346.6 million price tag associated with rulemaking alone.

RECLAIM was developed under some unique constraints. First, statutory and legal authority for mobile source control is shared with the California Air Resources Board (CARB). This agency largely develops statewide initiatives, which then have to be either tailored to the region's air management plant or factored in. RECLAIM was also developed with a variety of regulatory and legal design constraints, some of which were self-imposed. For example, the district prohibited the use of banking as an element in its program design. At the same time, the district was required to develop sunset provisions for the program, largely in response to constituent fears that the program wouldn't function. For the business community, this fear took the special form of concern over credit supplies and cost. As a result, price levels were set that would trigger abandonment of the emissions trading market as the primary compliance strategy. The district was prevented from using banking as the primary tool to assure adequate supplies as well as to insure against price spikes during compliance periods.

Recognizing the need for a market design change to deal with these two problems and attempting to build support among skeptical business people, the district hired a team of consultants who developed a strategy of overlapping permit issuance periods (Carlson and Scholtz 1994). This approach solves the problem of potential end-of-the-period price spiking, as it enables emitters to draw on RTCs of the next cycle rather than increasing the demand for RTCs of the current cycle. It does not solve the problem of emission spiking, which, of course, is the ostensible purpose of the banking ban. Emission spiking is a concern because the overlapping issuance periods allow firms to tap next-cycle RTCs to cover the current-cycle emissions. There is no assurance under this approach that conditions of peak demand will not exacerbate rather than alleviate ozone exceedances. This is a particular problem for utilities whose NO_x

emissions increase with peak electricity demands associated with weather conditions that exactly favor ozone formation.

In this regard, the approach introduces more administrative complexity than banking with no added performance benefit. Most emissions trading paradigms are based on the premise of ex post crediting for reductions and no reverse banking; that is, future emission entitlements cannot be turned into current emission increases. In fact, problems of inadequate attention to the environmental implications of trading system design and practice have more often impeded rather than aided the development of such markets.[12]

Further, in regions without the substantial trading experience that characterizes the Los Angeles region, sources are much more likely to fear that an effective market will not emerge. As a result, banking offers sources the encouragment to invest in overcontrol by allowing them the option of preserving the financial value of that investment while waiting for the market to emerge. In the absence of such encouragement, emissions will continue up the stack and be implicitly banked at the expense of the environment.

Chicago Nonattainment Area

Illinois has been a hotbed of emissions trading activity. The state has developed a conceptual framework for an emissions trading program that was based on an explicit areawide emissions cap for NO_x. At the eleventh hour, however, regional air quality modeling undertaken by the Lake Air Directors' Consortium (LADCO) indicated the need to change gears. The modeling showed that VOCs, not NO_x, reductions were required to meet the ozone standard within the Chicago metropolitan region. Illinois tabled its NO_x trading rule and developed a VOC reduction program that used emissions trading as its centerpiece in a strategy that ultimately followed the earlier NO_x emissions cap approach (IEPA 1995).

Illinois' new emission trading program first considered a two-phased approach. The first phase was to be structured like a traditional regulatory control program focused on the demonstration of reasonable further progress by 1999. During the period 1996–2000, sources could opt into the allotment process and begin both banking and trading by accepting a baseline. During the second phase, sources would be required to comply with a cap-and-trade model replete with allowance and true-up periods. An alternative was to

[12]See, for example, Liroff (1986).

drop the opt-in period so that all sources would enter the market in 1999. This latter approach was favored as it would simplify administration of the program.

Illinois developed both the NO_x and what came to be called volatile organic material (VOM) emissions trading programs by securing the input and participation of environmental groups, business, and representatives of groups advocating different trading models.[13] The plan represents a consensus approach that appears to have one of the highest potentials for success.

The development of both a NO_x and VOM trading program was fortuitous both for Illinois and the future course of ozone control programs. After the focus shifted to VOM, continued modeling studies indicated the importance of ozone transport into the nonattainment region for the magnitude of VOM reductions required within the region. This insight led to the prospect of an upwind NO_x trading program to reduce ozone inflows at the boundary of the nonattainment region. This insight, extending as it does beyond state boundaries, is the underlying vision of the 37-state Ozone Transport Assessment Group (OTAG), which is currently developing recommendations for NO_x transport and management.

Northeast States/NESCAUM and the OTC

The Northeast States for Coordinated Air Use and Management (NESCAUM) consist of Connecticut, Rhode Island, Maine, Massachusetts, New Hampshire, New Jersey, New York, and Vermont. This group has been extremely active in emissions trading and has joined forces with other states in the Ozone Transport Commission (OTC). NESCAUM is currently implementing two federal grants concerning emissions trading.

The first grant is for the development of an NO_x cap proposal in the Northeast. This project has been taken up by the OTC and is now in the form of an MOU (memo of understanding) among the states in the OTC. While many suspect that this MOU is in precarious condition, the OTC ratified a "model" allowance-trading rule designed to implement the MOU, and most OTC states have begun the rule-making process needed to adopt the model rule. The policy proposes to cap NO_x emissions at utility and other utility-sized boilers to create a regional trading program through the use of

[13]The Illinois EPA's VOM trading design team included representatives from the IEPA, Environmental Defense Fund, Corn Products, Coopers & Lybrand, Caterpillar, Abbot Laboratories, Commonwealth Edison, and Amoco Oil.

emission allowances. The OTC program is strikingly similar to the Title IV program for SO_2, mirroring virtually every design feature of the latter. In fact, the OTC states and the U.S. EPA are in the process of developing an agreement under which the EPA's current system for tracking SO_2 allowances and emissions would be adapted and applied to the tracking of the NO_x emissions and allowances of the sources affected by the OTC program. Like the SO_2 program, the OTC NO_x program both imposes a cap and permits intertemporal banking. The OTC program, however, imposes a restriction on the withdrawal of NO_x allowances from source's banks by requiring a two-for-one offset whenever the OTC-regionwide bank exceeds 10% of the cap and an individual source seeks to withdraw allowances from its own bank past a certain point. This provision reflects the need to address the contemporary, rather than cumulative, role NO_x plays in ozone formation.

The second grant to NESCAUM is for an emissions trading demonstration project. NESCAUM has coalesced a group of industries, environmentalists, interest groups, and regulators to demonstrate various emission reduction methodologies and to explore options for using credits. During the first two summers of the project (1992 and 1993), more than 5000 tons of NO_x and VOC (combined) were removed from the air by several means, including demand-side management (DSM), selective noncatalytic reduction (SNCR), fuel switching, and process changes. Through the NESCAUM process, involved parties had the chance to test, document, and work through the reduction side of an emissions trade. During the third summer, the project moved into the "use" end of the equation. Several model trades were identified and worked through. This process enabled the various parties to determine where policy problems would potentially arise and helped to find measures to address these problems. The NESCAUM trades have provided the template for the EPA's proposed DER program.

Ozone Transport Commission (OTC)

Interstate Conformity

The ozone transport region was officially created in Section 176a of the Clean Air Act and consists of the NESCAUM states as well as Delaware, Maryland, Pennsylvania, and the District of Columbia. Its creation represents a legal acknowledgment of the importance of transport, i.e., transboundary pollutant flows, in ozone control.

States in the OTC face certain requirements beyond those required as part of a nonattainment area. The member states formed an organization, the Ozone Transport Commission (OTC), to examine interstate transport issues. The OTC has communicated with EPA regarding interstate issues and is coordinating with all member states to foster regulatory consistency between the states on issues relating to ozone transport. As the preceding section illustrated, the OTC with the Midwest, is at the cutting edge of market-based policy development. Few states are large enough to pretend to encompass the geographic air sheds that govern air pollution problems. This is, in fact, the rationale for the creation of the OTC. Traditionally, ozone transport has either been ignored by states in the formulation of attainment plans or been left to resolution by EPA in its review of individual state implementation plans (SIPs).

For these reasons, the OTC is moving to create an interstate trading system for NO_x—a system that could rival in size and value the SO_2 allowance market. The OTC's model rule development was a response to the primary challenge facing the OTC to develop conformity rules between states. This issue will increase in importance as the OTC attempts to coordinate policies with the Midwest. This potential integration is currently being explored through the OTAG process. In this setting, it is expected that transport conditions will be recognized throughout the eastern United States. The expansion of the relevant NO_x control region will relieve internal OTC political pressures and allow, as well, for more rational integration of electricity and emissions markets.

Mobile Sources

In 1994, the OTC petitioned EPA to impose the California car program throughout the region to address one aspect of the ozone problem plaguing the citizens of the Northeast. The petition intensified the battle over the cost-effectiveness of the California program, threatening to undermine the credibility of attainment plans throughout the region. The ensuing heated debate focused only on the technical and economic feasibility of the California car program in the Northeast as is typical of the traditional command-and-control regulatory process. Little attention was paid to either cost or effectiveness.

In response, at least one market-based policy has been proposed that would put a safety net under both the environment and the economy by simultaneously delivering both an increased certainty of emissions reductions and the potential for cost savings (Dudek and Goffman 1994).

The essential ingredients of this proposal are:

- The transformation of the California car program in the Northeast into specific mandated quantities of emissions reductions.
- The legally enforceable assignment of those emissions reduction responsibilities to the automobile manufacturers.

The proposal put forward by the Environmental Defense Fund (EDF) would create a vehicle emissions reduction responsibility (VERR) by establishing a bubble over those elements of the California car program in dispute for application in the Northeast. If onboard vehicle technologies turn out to be as cost-effective as their advocates claim, then some mix of certified low-emitting vehicles (TLEVs, LEVs, ULEVs, and ZEVs) will be introduced by automobile manufacturers in order to meet their individual VERR. If, on the other hand, the tailpipe emission limits imposed under the California car program turn out to be a technologic and economic stretch, then the automobile manufacturers will have the option of meeting their emissions reductions obligations by producing legally equivalent emissions reductions from other sources within the region.

Economic analysis of the California car program applied in the Northeast indicates that the region could spend between $2 billion and $23 billion in present value terms on vehicle emissions control for the period 1996–2003. The Environmental Defense Fund's flexible compliance strategy's costs are estimated to range from $5.6 to $6.8 billion, based on the costs of acquiring compensating emissions reductions. However, since EDF's proposal preserves the option of producing "the California car," if vehicle emissions control proves to be cost-effective, then the region would spend only the $2 billion. The effect of EDF's vehicle emissions reduction responsibility proposal then is to put an upper limit on compliance costs. More importantly, adoption of EDF's proposal would guarantee the production of the required emissions reductions irrespective of the method of compliance or the future evolution of the technology and economy. If adopted, this proposal could catalyze the development of a broader regionwide trading system for both NO_x and VOCs while avoiding the continued programmatic separation of pollution sources that characterizes most state programs. At present, EPA and the Big Three have agreed to a more expensive and less environmentally protective alternative—a 49-state car that improves on present Tier I requirements one year sooner than EPA could go to Tier II vehicle requirements.

SYNTHESIS AND CONCLUSIONS

Synthesis

Emissions trading systems have been developed both to reduce the tension between economic growth and air quality and to increase the performance of environmental programs. The ability of emissions trading programs to achieve the full measure of these benefits, however, depends critically on the design of the program. The cap-and-trade paradigm, exemplified by the SO_2 program, with its rigorous compliance scheme and the fluidity of allowances as currency in an emissions reduction market, comes closest to creating the conditions under which these dynamics can be generated.

The role of banking in the acid rain program illustrates this clearly. Sensitive ecosystems have been damaged by the cumulative efforts of the SO_2-related deposition. At the same time, the sooner SO_2 reductions are made, the sooner affected systems can begin to recover. Accordingly, Congress permitted unlimited banking and intertemporal trading, which would yield no fewer cumulative reductions than would be achieved in the absence of such trading. Moreover, such banking offers utilities a rationale for early reductions. The EPA has projected that utilities are responding to this opportunity by making 40% more reductions than required in the first five years of the program.[14] This is strong evidence that, in addition to providing the opportunity for banking, the key design elements of the acid rain program have created powerful incentives for early reductions.

Specifically, the cap on annual emissions, together with the certainty for potential purchasers that early reductions can be used to meet compliance under the cap, signals utilities that early extra reductions can be a valuable asset, financially justifying investments in early reductions. The apparent success of the acid rain program has persuaded at least two jurisdictions to apply the lessons of the SO_2 trading program to solving the problem of ozone nonattainment. While ground-level ozone formation is regulated as an acute, rather than cumulative, threat to human health, both the Illinois EPA (IEPA), which is targeting VOCs, and the Ozone Transport Commission (OTC), which is targeting NO_x, are developing programs for reducing ozone precursors that not only use the cap-and-trade model

---

[14]Testimony of Mary Nichols, assistant administrator for air and radiation, U.S. EPA, in hearings before the Subcommittee on Energy and Power, Committee on Energy and Commerce, on Title IV of the Clean Air Act, October 5, 1994.

but permit banking, albeit for only one year in the cap of the IEPA and with a restriction on the use of banked allowances in the case of the OTC.

By using this approach, the designers of these programs are seeking to capture the benefits of cleaner air by early overcontrol as well as by cost-savings and innovations created by the cap-and-trade design of the acid rain program. This policy choice reflects an explicit determination that the potential for achieving these benefits outweighs the risk that early, banked reductions will be used to offset unacceptable emissions increases in future ozone sessions. This determination is supported by the continuous force of incentives, lasting for the life of the programs, to produce extra reductions for future cost-effective compliance or sale in the emissions reduction market. Consequently, scenarios under which a preponderance of sources will find that economic conditions compel them to empty their banks at precisely the same moment are highly unlikely.

Conversely, if the IEPA and OTC proposals did not include banking, then the likelihood of achieving the benefits of early overcontrol investment in banked credits and accelerated reductions would be vastly diminished. Unless sources are confident that an emissions trading market will be robust—which is often in doubt in the early years of a program, when decisions to engage in overcontrol are most critical—they will shy away from making aggressive investments early on. Since many overcontrol strategies are "lumpy" and not gradual and often result in sources locking into a specific compliance approach, the greater-than-required reduction benefits persist over a long term. These very same characteristics, at the same time, demand that clear-cut economic incentives be present and strong enough to justify such investments in the first place. Limited year-to-year banking in these designs represents a balance between market stimulation and direct contemporaneous emissions reductions benefits.

Understanding the Alternative Models

The efforts of the IEPA, the OTC and the California South Coast Air Quality Management District (SCAQMD) through its RECLAIM program to apply at least the core design features of the cap-and-trade model to the control of ozone precursors emphasizes that air quality program designers have the opportunity to select from a variety of templates in using emissions trading in the design of their programs. Until the efforts of these three agencies to develop alternative market models, it was generally believed that the emission reduction credit (ERC) model was the only one that could apply to the control of ozone precursors. This assumption was further challenged when the U.S. EPA began discussions concerning an "open-market" trading rule intended to apply primarily to ozone precursors.

If the goal of emissions trading is not just to cut costs but also to achieve fundamental policy reform that can yield superior environmental performance, then a clear hierarchy, led by the cap-and-trade model, emerges among these three alternatives. The regulator's challenge is to identify the proper integration of these three models.

Under the Clean Air Act, ozone formation is regulated as an acute, daily problem rather than a total loading problem. Nonetheless, a program for reducing the pollutants NO_x and VOC's, which are ozone precursors, by using the cap-and-trade model still allows ozone attainment programs to improve their effectiveness while capturing the benefits of the cap-and-trade model. Ultimately, these benefits are just as critical, if not more critical, to attaining the ozone standard as they are to combating acid rain. To do so, ozone attainment plans, as well as programs that achieve the reasonable further progress toward attainment as required under the Clean Air Act, must have the effect of capping actual emissions of NO_x and/or VOCs. In addition, these programs must be affordable and, where substantial reductions are needed, they must foster early reductions and continuous innovation; hence, the embrace by SCAQMD, IEPA, and OTC of the cap-and-trade model to achieve substantial ozone precursor reductions.

Well-designed ERC-based approaches can claim to deliver to those individual sources that elect to engage in trading some of the benefits of a comprehensive cap-and-trade system. In most such programs, sources—at least those supplying ERCs—are effectively subject to a cap on their actual emissions as a result of the certification process. In addition, ERCs, once certified, can be exchanged in transactions that require relatively little regulatory intervention. Nevertheless, ERC trading is based on existing CAC regulations, and sources that do not trade are not subject to a limitation on total emissions or to the incentives for cost savings, overcontrol, and innovation in the way that all sources in cap-and-trade systems are. More importantly, programs based on ERCs are optional. They exist at the whim of states and are utilized at the discretion of sources irrespective of environmental exigencies.

The EPA's "open market" (DER proposal) is like the ERC system in that it offers a flexible compliance option to sources meeting existing CAC regulations. Because open-market trading can include "discrete," or one-time, reductions, even sources that do trade are not subject to continuous caps on their actual emissions.[15] For these reasons alone, the open-market approach scores even lower on the test of true policy reform. In addition, the approaches to the certification of excess reductions that the EPA is considering may create either high levels of uncertainty as to the value of such reductions or such low levels of credibility with respect to compliance that open-market programs might not even deliver the basic required reductions, let alone the sophisticated economic dynamics of emissions reduction markets.

Specifically, the open-market guidelines may permit one of two approaches. Under the first, firms would transact excess "discrete" reductions. These reductions would be certified as usable for compliance only when the purchaser submitted them to state regulators to offset the purchaser's emissions for purposes of meeting compliance. Under such a scenario, the uncertainties with respect to the usability of the reductions may render them nearly valueless to potential buyers and, therefore, to potential investors in overcontrol. State regulators could face a corresponding risk that political realities would prevent them from thoroughly scrutinizing such reductions to the extent that economically or politically significant sources relied on their purchase to meet compliance. In that case, the integrity of the underlying compliance scheme could be eroded, with the result that air quality programs would fail to meet their intended emissions reduction objectives. Areas in which substantial reductions are needed to attain the health-based ozone standard cannot afford the loss of reductions resulting from weakened compliance.

To cure the problem of regulatory uncertainty inherent in the open market's buyer beware, post hoc certification approach, some have proposed a second approach, that of reducing public regulatory oversight to the vanishing point. Under this approach, firms would transact emissions trades and be subject only to occasional, random audits by state environmental agencies. Unfortunately, this proposal does not necessarily relieve firms of expensive transaction costs, but it does threaten to undermine severely the compliance credibility of the underlying emissions reduction programs. However occasional regulatory audits might be, conscientious firms seeking to acquire

---

[15]Of course, sources are limited by an implicit cap on emissions, which is the allowable emissions rate at maximum throughput. The present permitted levels can be viewed as only temporary in the face of impending nonattainment plans, and yet the DER proposal would treat these limits as a permanent cap and offer the flexibility of unlimited banking normally reserved for fixed caps.

emissions reductions would be compelled to maintain reliable records of selling firms' reductions and even to obtain indemnification from sellers. At the same time, for less scrupulous firms, the likely infrequency of audits would make indifference to the quality, or even the existence, of purchased reductions an economically viable option. For these reasons, the failure to attain environmental goals or the abuse of the program's flexibility could result in a public loss of confidence in emissions trading in general.

The contrast provided by a cap-and-trade program dramatizes the problems inherent in the DER approach. With allowance trading, firms can exchange emissions reductions without having to bear any costs related to ensuring the environmental performance of the selling firm and without the need for indemnification related to environmental nonperformance. In fact, there are virtually no transaction costs related to regulatory requirements imposed on trading itself.[16] Firms can simply transfer emissions allowances through simple contracts, without subjecting the transfers to regulatory intervention. At the same time, the credibility of the underlying emissions reduction compliance scheme remains wholly uncompromised.

The proposed open-market alternative, however, transfers to the private sector certification and indemnification costs that either are not present in cap-and-trade designs or are built into the components of the public compliance structure and are thus borne by the regulatory process. It turns out, then, that not only are there no cost savings to be gained by moving from the rigorous compliance scheme of a cap-and-trade system to "privatized" compliance under the open-market system but net transaction costs to sources may actually be higher under the latter approach. Meanwhile, these additional private sector costs do not yield greater environmental benefits but, rather, are linked with substantially diminished assurance that emissions reductions will be achieved.

Once again, the role of intertemporal banking is highly illustrative of the extreme limitations of the open-market approach. As currently formulated, the open-market proposal would allow sources to use discrete emission reductions from any year subsequent to their creation (i.e., their legal recognition at the time of the compliance equivalence request), even when the pollutants involved are ozone precursors. For example, emissions reductions produced decades

---

[16]Of course, for publicly regulated utilities subject to regulatory oversight of compliance expenditures, this remains a problem that has chilled market development. For additional details, see U.S. General Accounting Office (1994).

ago could be presented for compliance against the present season's emissions. In this case, however, the justification for the risk of moving emissions increases into future years is absent because the open market lacks the critical ingredients needed to stimulate early investment in overcontrol. The lack of a cap, together with the uncertainties about the regulatory acceptability of specific DERs, the potential high transactions cost, and even the potential erosion of the compliance program itself, all call into question the economic value of creating excess reductions. Firms would be likely to have little confidence that investments in overcontrol would yield additional economic return. Without a serious expectation that those investments would occur, the rationale for banking as a strategy for enhancing the environmental performance of a program, which can be quite powerful in a cap-and-trade program, cannot be sustained in an open-market system.

The Emissions Trading Hierarchy

The powerful case for the dramatic superiority of cap-and-trade systems suggests that regulators, environmentalists, and the business community should seek the widest possible application of such systems to air quality problems. Clearly, in areas where substantial reductions are needed to attain the ozone standard, the need for cost savings, innovation, and certainty in achieving the emissions reduction target demand such an approach. Existing information and the ready feasibility of monitoring or otherwise quantifying emissions from many categories of sources bring the establishment of such programs readily within reach, as SCAQMD, IEPA, and OTC have already demonstrated to varying degrees.

At the same time, however, regulators are presented with a wide variety of circumstances extending far beyond the one just described. Some areas need only maintain current air quality levels. Some categories of sources do not lend themselves, at least at present, to being regulated through the allocation of explicit emissions limitations and to precise emissions monitoring. For some source categories, establishing a cap-and-trade system may consume time during which early reduction opportunities might be lost. These and similar scenarios demand that the policy designer's toolbox contain more than just the cap-and-trade model. The challenge, then, is to identify when ERC or even open-market systems might be appropriate and how they can stimulate the development or expansion of, and be integrated with, air quality programs that rely on cap-and-trade models as their centerpiece.

In most cases, ERC-based systems offer the best complement to, or developmental precursor for, cap-and-trade systems. One of the essential elements of this approach is that transacting sources are explicitly or implicitly subject to a total emissions cap as part of the ERC certification process, which occurs at the inception of the transaction. The de facto source cap and the reliability of the ERC currency created by the certification process make ERC trading fundamentally compatible with cap-and-trade systems. Typically, the criticism of ERCs is that the initial certification process imposes high transaction costs on sources. The open-market concept is intended, in part, to remedy that. In fact, with the increasing availability of more reliable and accurate monitoring and quantification techniques, a more modest, but ultimately more effective, solution may be at hand. In lieu of traditional, burdensome up-front certification processes, transacting sources could simply agree to adopt continuous emissions monitoring or the equivalent. This would give regulators a more reliable alternative to current methods of calculation and projection of the expected availability of the selling sources' surplus reductions and the appropriate use by purchasing sources.

This last point is especially important in light of the argument that the open market is a superior strategy for developing the information necessary about the actual emissions of sources necessary to effectuate the allocation of emissions limitation responsibility that is, in turn, the cornerstone of cap-and-trade systems. With increasing reliance on monitoring and/or quantification in place of up-front certifications, ERC systems can lessen initial regulatory transaction costs while increasing the amount of emissions information available about sources for the development of more comprehensive cap-and-trade systems. Because ERC systems already possess important elements of compatibility with comprehensive cap-and-trade systems, this approach can enhance their role in accelerating the evolution of cap-and-trade programs for sources not immediately agreeable to the creation of such programs without an intervening stage.

If open-market programs bear out the claims of their proponents, they, too, could offer a key component of the transition to cap-and-trade systems. Again, that component is the information about individual source and categorywide emissions. Both the post hoc certification and the random audit versions of the open-market model do put a premium on sources' generation of reliable emissions information. However, neither of these provisions would address the problem of crediting and transacting the difference between allowable and actual emissions—differences that, if accepted by regulators, could result only in net emissions increases. Even if the success of such programs in this regard could be assumed, notwithstanding the profound doubts about the overall credibility of these approaches, the dissimilarities between cap-based programs and the open market at present make it very unclear how to accomplish an integrated transition from the latter to the former. For example, the open-market model permits trading in "discrete" reductions, rejecting the explicit or de facto source caps created in ERC transactions. As a result, it is difficult to fathom how open-market reductions can be integrated into the emissions reduction "currency" of cap-and-trade systems. Until this is resolved, the open market's role in transition to cap-and-trade programs seems limited to the vanishing point.

This problem is compounded by the emerging need to address the interstate transport of ozone and NO_x. Emissions reduction markets can link sources across state borders and, among other things, provide interstate cost sharing, but only if those sources generate compatible emission reduction currencies. At present, there seems to be little that open-market models can contribute to the creation of such currencies.

The most persuasive role for open-market trading is to be found in two areas, where achieving a precise emissions limitation target may be substantially less critical. First, where areas have attained the ozone standard and communities must keep total emissions within a certain range, open-market trades may increase sources' flexibility without exceeding the air shed's tolerance band. Second, where states have adopted specific CAC requirements such as RACT as a transitional step to achieving attainment, open-market trading may be acceptable as an alternative compliance mechanism that is superior to regulatory grants of variances, extensions, and alternative emissions limits. In these cases, again, hitting a precise total emissions target on which reaching attainment may depend is not the objective of strategies like RACT. Instead, these programs are called on to initiate increments of progress toward attainment, in effect, setting the stage for more aggressive strategies that must demonstrate the achievement of emissions limitations of sufficient stringency to ensure attainment. If the application of open-market trading is confined to these roles, then it can take a constructive place in the hierarchy of emissions trading models.[17]

[17]Some have questioned the legality of applying open-market trading to RACT without formally amending federal RACT regulations. Moreover, trading under RACT could be effectuated through a cap-and-trade approach under which reductions expected as a result of RACT programs are codified as an explicit cap and allocated to affected sources in the form of tradable allowances.

Conclusions

Each of the basic emissions trading systems can produce both economic and environmental benefits. Care must be taken, however, to match the program's design and application to the underlying physical nature of the environmental problem in both spatial and temporal dimensions. Active market development has been hampered by uncertainty on the part of buyers that their purchases will be honored as fully equivalent compliance options by the various regulatory authorities overseeing transactions. Each system provides a different level of assurance that environmental goals and increased environmental performance will be achieved. Each system creates a different set of incentives for program participants, and each distributes differently both the risks and rewards of emissions trading. Each system provides encouragment for early reductions, but the stimulus to innovation is closely linked to the ability of risk takers to recover value for their investments. Regulators need to pick the emissions trading tool that best matches the nature of the underlying environmental problem. Emissions trading systems work best when all stakeholders believe that transactions involve "excess reductions" between sources to voluntarily lower and redistribute the costs of compliance while guaranteeing that the overarching environmental goal will be met.

Despite almost 20 years of experimentation with markets for environmental commodities, these markets remain limited in scope and application. Even the SO_2 allowance market, the most ambitious and successful example to date, continues to attract critics. Emissions trading continues to undergo exploration and development in the search for more cost-effective tools to solve additional complex and diffuse environmental problems. In the process, it is hoped that stakeholders will accumulate enough experience to strike the necessary balance among flexibility, acountability, and performance in order to give the public the assurance that market forces can indeed be harnessed for environmental protection.

REFERENCES

Carlson, Dale, A., and Anne M. Scholtz. 1994. "RECLAIM: Lessons From Southern California for Environmental Markets." *Journal of Environmental Law and Practice.* 1(4):15–26.

Dudek, Daniel J. 1989. "Emissions Trading: Environmental Perestroika or Flimflam?" *The Electricity Journal* 2(9):32–43.

Dudek, Daniel J. 1993. "Incentives and the Car." in *Cost-Effective Control of Urban Smog*, edited by R. F. Kosobud, W. A. Testa, and D. A. Hanson. Chicago: Federal Reserve Bank of Chicago. November.

Dudek, Daniel J., and Joseph Goffman. 1994. "Flexibility for Responsibility: Focusing on Reductions Rather Than Technology in Vehicle Emissions Control." New York: Environmental Defense Fund. June.

Dudek, Daniel J. and John Palmissano. 1988. "Emissions Trading: Why Is This Thoroughbred Hobbled?" *Columbia Journal of Environmental Law.* 213(2):217–256. October.

Dudek, Daniel J. and James T. B. Tripp. 1989. "Institutional Guidelines for Designing Successful Transferable Rights Programs," *Yale Journal on Regulation* 6:381–403. June.

Dudek, Daniel J., Joseph Goffman, and Tom Walton. 1992. "Mobile Emission Reduction Crediting: A Clean Air Act Economic Incentive Policy Proposal for retiring High-Emitting Vehicles." Environmental Defense Fund and General Motors Corporation publication. New York: Environmental Defense Fund, 16th Floor, 257 Park Avenue South, New York, N.Y. 10010. October.

Dudek, Daniel J., Zbigniew Kulczynski, and Tomas Zylicz. 1993. "Implementing Tradable Rights in Poland: A Case Study of Chorzow." Conference Proceedings of the Third Annual Meeting of the European Association of Environmental and Resource Economists, *Economics and Environment No 11*. Amsterdam, Holland. November.

Environmental Defense Fund. 1995. *Environmental Compliance Alert.* II, 5. Washington, D.C. February.

Illinois Environmental Protection Agency. 1995. *Final Proposal for VOM Emissions Trading System.* Springfield. March.

Liroff, Richard. 1986. "Reforming Air Pollution Regulation: The Toil and Trouble of EPA's Bubble Policy." Washington, D.C. The Conservation Foundation.

U.S. Environmental Protection Agency. 1993. *Federal Register.* II (5): p. 11110. February 23.

U.S. General Accounting Office. 1994. "Allowance Trading Offers an Opportunity to Reduce Emissions at Less Cost." Washington, D.C.: GAO/RCED-95-30. December.

Wirth, Timothy E., and John Heinz. 1988. "Project 88—Harnessing Market Forces to Protect the Environment: Initiatives for the New President." Environmental Policy Institute. Washington, D.C. December.

APPENDIX: THE 1990 CLEAN AIR ACT AMENDMENTS AND OZONE REQUIREMENTS

Title I of the 1990 Clean Air Act Amendments (CAAA) requires states to develop state implementation plans (SIPs) indicating how they will meet the National Ambient Air Quality Standards (NAAQS) for the criteria pollutants: ozone (O_3), carbon monoxide (CO), particulate matter (PM10), sulfur dioxide (SO_2), and lead (Pb).

The SIPs must also indicate how states plan to maintain air quality once the NAAQS are met. The CAAA includes a series of command-and-control requirements based on the attainment classification of an area for a specific criteria pollutant. These requirements are the foundation on which states must build (or remodel) their air pollution control programs. The requirements for the reduction of ozone are the most detailed in Title I, followed by those for carbon monoxide. As for the other criteria pollutants, Title I lays out the criteria for designating an area in nonattainment and establishes a deadline for compliance.

The requirements for ozone are broken out by nonattainment category: marginal, moderate, serious, severe, and extreme. Marginal nonattainment areas are the least polluted. The requirements that apply to these areas also apply to all areas with a higher classification. Likewise, all requirements that apply to a nonattainment area also apply to those with higher classifications.

The requirements for *marginal areas and above* include:

- The need for a detailed emission inventory.
- The application of RACT to all major stationary sources (defined as emitting at least 100 tons per year).
- Basic inspection and maintenance (I & M) standards.
- Submittal of a periodic inventory every three years.
- Adoption of an emission statements program.
- New source review (NSR) offsets at a minimum ratio of 1.1:1.
- The option to adopt a reformulated gasoline program and the sanction of being automatically bumped into the next-higher classification for failure to comply with the NAAQS by the prescribed date.

Moderate areas and above share the above requirements as well as the following added responsibilities:

- Fifteen percent of inventoried emissions must be reduced by 1996.
- There must be an annual demonstration of compliance.
- The SIP must contain contingency measures that automatically kick in if the state's demonstration fails to show compliance with the NAAQS.
- The state must submit tracking plans to document its progress toward attainment.
- Gasoline stations handling more than 10,000 gallons per month

must install Stage II vapor recovery devices, and the NSR offset ratio increases to 1.15:1.

Serious areas and above face these additional requirements:

- The definition of major source climbs to all sources that emit at least 50 tons per year.
- The NSR ratio climbs to 1.2:1, and modifications can exempt out of the requirement if they can meet an internal offset ratio of 1.3:1.
- The state must implement a system for enhanced monitoring of the ambient air.
- The attainment demonstration in 1994 must show compliance using photochemical grid modeling.
- The reasonable further progress (RFP) demonstration must include a minimum annual reduction of 3% above and beyond the original 15% requirement.
- Enhanced I & M requirements with a $450 waiver fee.
- A clean fleet vehicle program for centrally fueled fleets.
- Transportation controls.
- Reformulated gasoline requirements.
- Long-term backstop provisions to prevent deterioration.

Severe areas and above must also:

- Redefine major sources as those that emit at least 25 tons per year.
- Increase the NSR ratio to at least 1.3:1.
- Develop transportation control measures to offset emissions increases from any increase in vehicle miles traveled.
- Require employee trip reduction programs of certain employers.

In addition, *extreme areas* must:

- Redefine major sources as those that emit at least 10 tons per year.
- Increase the NSR ratio to 1.5:1.
- Require clean fuels for boilers.
- Institute traffic control measures during heavy traffic hours.

Finally, those areas designated as part of an ozone transport region (so far there is only one, comprising the states of Connecticut,

Delaware, Maine, Maryland, Massachusetts, New Hampshire, New Jersey, New York, Pennsylvania, Rhode Island, Vermont and the District of Columbia), regardless of individual attainment classifications, must:

- Institute an enhanced I & M program.
- Apply the NSR offset requirements to all new sources or modifications to existing sources where the increase in emissions is at least 50 tons per year.
- Require Stage II Vapor Recovery Requirements at gasoline stations with throughput greater than 10,000 gallons per month.

The 1990 CAAA allows, and even encourages, the use of economic incentives as a strategy to implement air pollution reduction requirements. Economic incentives are supported as a means of increasing the efficiency of pollution control programs while, at the same time, reducing their cost. However, many states are finding it difficult to construct an incentive-based pollution control program on top of existing rulemaking or CAC-based regulatory frameworks, given the preceding requirements.

This transition is particularly difficult for ambient ozone and criteria pollutants because the 1990 CAA does not clearly endorse and/or administratively support emissions trading—or any other form of economic incentive—as the preferred strategy for achieving clean air. States wishing to develop programs must conform to several different guidelines and policies, including the 1986 Emissions Trading Policy (ETP), the 1990 CAAA, the 1993 Economic Incentive Rule (EIR), and the 1993 Guidance on the Generation of Mobile Source Emission Reduction Credits (MERCs). These guidance materials merely establish parameters for trading regulations; it is up to individual states to craft their own regulations in such a way as to pass muster at EPA during the SIP review process and, more critically, to encourage emissions sources to participate in the trading program.

2.7

❑

SLIPPAGE FACTORS IN EMISSIONS TRADING

George S. Tolley and Brian K. Edwards

INTRODUCTION

❑ The favorable news about emissions trading is that it gives incentives to find least-cost ways of achieving environmental aims. This is no small matter in the present era, when increasingly expensive measures are required to make further environmental improvements under the Clean Air Act Amendments. Emissions trading opens up the possibility for some firms to buy emission reduction credits (ERCs) from others for less than their on-site costs of reducing emissions. Other firms, with on-site emissions reduction costs lower than the price they can receive from selling credits, have incentives to keep their emissions low and sell the excess of their allowed emissions over actual emissions to those whose emissions reduction costs are high. In this way, the emissions reduction job is transferred from high-cost to low-cost emission reducers. If firms are allowed flexibility in the technologies they use to reduce emissions rather than having their technology commanded and controlled by government directives, market incentives will exist to develop lower-cost technologies for reducing emissions.

While the move toward the use of markets is certainly to be welcomed and encouraged, a number of factors could prevent emissions trading from achieving all the gains hoped for. These factors, which we call *slippage factors*, have been relatively neglected in analyses of emissions trading.

Important studies have estimated the potential gains from emissions trading. For example, in planning before the fact for its

187

Regional Clean Air Incentives Market (RECLAIM) emissions trading program, the South Coast Air Quality Management District (1993) estimated that a system of markets for SO_x and NO_x emissions reductions could reduce control costs in the Los Angeles area by as much as 50%. The district's audit report (1996), conducted after the program was put into effect, contained information on trading prices and the amounts of trading but did not estimate cost savings. As Dr. Anupom Ganguli from the district pointed out in conversation, it is difficult to measure savings in practice. In addition to the problem of obtaining accurate information on control costs borne by emitters, the control costs they would bear in the absence of emissions trading is a hypothetical that could require a significant research effort to estimate. In the absence of direct estimates of realized cost savings, the volume of emissions trading—which generally appears to remain small relative to total emissions—and the lower than expected prices of emissions rights in the various places around the country where emissions trading has been introduced are suggestive of the very real existence of slippage factors.

A key question is how to cope with so-called transaction costs, which will inhibit realizations of gains that would occur in a frictionless situation. The term *transaction costs* may give the impression that some minor impediments exist that will somehow not be important or that can be left to others to worry about. Rather, to be realistic about what can be achieved from emissions trading and, more important, to design systems that will be as effective as possible, attention to these impediments and how to cope with them should be the name of the game. We mentioned some of the problems briefly in Tolley et al. (1993). The following, more comprehensive remarks suggest 14 points that need to be faced up to.

Regulation Must Be Concise Enough to Allow Accurate and Simple Recordkeeping and Monitoring

The knowledge required, the trouble, and the use of valuable managerial time needed to comply even with seemingly straightforward command-and-control regulations, such as reasonable available control technology (RACT) or other technology-based standards, is easily underestimated. Emissions trading adds another level of complication in terms of keeping closer track than is now required of allowed and actual emissions, drawing up property rights documents with legal counsel, recording trades, and conforming to additional regulatory minutiae. Ignoring these complications or failing to keep them to a minimum could greatly restrict participation in trading.

Many Small Polluters Are a Weakness

The admonition to keep trading regulations simple is more important the more prevalent are small-business emitters. Managers of small businesses are even more likely than managers of large businesses to be skeptical of the unknown. They do not have the time or staff to be bothered with newfangled arrangements and will choose simpler compliance methods, if available. Managers of small businesses are likely to view the costs of obtaining the information and know-how necessary to participate in the market as greater than any expected benefits from it. These businesses must be kept in mind in the interests of wide trading participation.

Determining the Parties Who Would Be Willing to Engage in Trades Is Difficult

The distribution of potential participants according to size of business is only one of a number of market structure characteristics important to the success of trading programs. Factors affecting the ease of control among industries and the strengths of economic incentives to participate are also important. Other things constant, diversity among firms is conducive to trading if it means that some firms have high costs of reducing emissions and others have low costs. If all firms were identical, they would all move to the same point of emissions reduction with the same costs. There would be nothing to be gained from trading. A needed prelude to instituting a trading market is to survey emitters to find what considerations will limit their particpation, meeting their concerns if possible.

Emissions in One Location Have Different Effects than Emissions in Another Location

The nature of pollution is that damages from emissions vary with distance from the source. Damages depend on concentrations of receptors subject to damages at various distances. No two emissions sources are likely to be identical in damages caused by a pound of pollutants. Emissions reductions in a rural area may reduce damages very little because there are relatively few receptors, whereas the same reduction in an urban area could result in large reductions in damages.

The important regulatory consideration of locational differences in damages needs to be recognized through degradation and augmentation factors usable in emissions trades. The reduction in damages from cutting back by a pound of pollutant in one area may fall short of the increase in damages when pollution expands by a pound in another area. Although there is no net change in emissions, environmental quality is worsened. Conversely, environmental quality is improved if emissions reductions in one area reduce damages to a greater degree than the corresponding increase in damages in another area.

A study by Steven Puller (1993) demonstrates, for the first time that we know of, how to implement this approach. Estimates of externality values for SO_2, NO_x, and particulates were developed for two areas in Illinois using the damages approach The Coffeen power plant operated by Central Illinois Public Service is downstate in an area with low population density. The Crawford plant operated by Commonwealth Edison is located in densely populated Chicago, where more people are affected.

Externality costs were found to be manyfold higher in Chicago than in the downstate rural area. An additional ton of SO_2 emitted in an urban area causes $5,245 in damages, while the same ton emitted in a rural area would cause only $16 in damages. For particulates and NO_x, the differences between urban and rural externality values are even more dramatic. While rural and urban differences in externality values will not always be this great, the Puller study demonstrates the importance of recognizing locational differences in environmental externality values. The approach is feasible. It has logic and professional opinion on its side, giving it a good chance to prevail in the regulatory process.

The possibility that has been emphasized here is to incorporate degradation or multiplication factors into emissions trading rules. Another possibility may be mentioned that might circumvent many of the problems of emissions trading. This would be to proceed directly to a *market in damages caused by emissions* rather than trading emissions as such. Sellers would not sell emission rights but, rather, would sell the right to the damage reductions from reducing emissions. Buyers would be willing to pay for the rights to cause an equivalent amount of damages up to their costs of avoiding these damages. The objection might be raised that the approach is impractical because it requires quantifying damages that are difficult to measure. But damage measures are becoming increasingly routine. The difficulties might be no more insurmountable than those connected with emissions trading.

Ozone Modeling Problems Will Add Uncertainty to Trades

If we ignore the differences that a pound of pollutant can make in one place as opposed to another, we run the danger of blunting the advantages of trading or even not getting any benefits at all, as mentioned earlier. At the same time, if we try to recognize the true differences in emissions effects, we run into modeling uncertainties. The measurement of emissions and the modeling of their effects on air quality are often treated as sacrosanct in the regulatory process. However, measuring emissions and ambient concentrations is subject to great scientific uncertainty. Results can be affected by the placement of air quality monitors and by temperature when readings are taken. For example, the Chicago metropolitan area is out of attainment because of readings from one monitor in Waukegan, Illinois. Such measurement problems exist in other areas of the country. Only if we are reasonably sure about the magnitude of degradation and augmentation factors to attach to emission trades, can we be safe in assuming that a trade will actually reduce the cost of environmental control by more than the benefits are reduced in view of degradation.

The operational significance of the air quality modeling problems may be that effective emissions trading markets are more feasible for inert pollutants such as SO_2 than for reactive pollutants such as hydrocarbons that lead to ozone because modeling of inert pollutants appears more reliable than modeling of reactive pollutants.

Hand in hand with air quality modeling complications is the consideration that property rights have to be simple enough to be well defined. If one is going to get into very fine distinctions as to the season, the time of day, and small, highly defined geographic details, the whole approach could fall down on the basis of requiring more complexity than is feasible. One of the greatest challenges yet to be dealt with is how to design emissions trading systems that take account of enough of the differences in damages caused by differences in time and place of emissions to be sure environmental improvements will result from the trading while remaining simple and straightforward enough to be tractable from a legal and administrative point of view.

Surveys Have Indicated That Many, If Not Most, Utilities Will Rely on Measures Other than Emissions Trading to Meet Air Quality Requirements

The hesitancy of utilities to enter emissions trading markets stems in part from a long-standing orientation to technology-based solutions on the part of utility corporate planning and environmental departments. These habits are not changed overnight. Another consideration is uncertainty as to what treatment of ERCs and revenues from their sale will be allowed by state regulatory commissions. Uncertainty as to this treatment puts a damper on what utilities would be willing to pay for the allowances and may be partly responsible for the low price range in which SO_2 allowances have traded. For these and related reasons, many utilities still perceive it to be in their best interests to purchase more expensive pollution abatement equipment rather than treating allowances like any other financial asset offered in the marketplace.

Installing Additional Pollution Abatement Equipment to Obtain Allowances to Sell to Another Polluter Imposes Technological Risks on the Seller

An emitter choosing to sell allowances and make up the difference by installing pollution control equipment assumes the technological risks of how well the equipment will perform in reducing emissions. The cost may be greater than anticipated.

The Possibility that More Stringent Regulation Will Be Imposed in the Future May Lead to Hoarding Emissions Credits, Making Them Less Available to the Market

The more stringent regulations in prospect under the Clean Air Act Amendments will require emissions reductions beyond those already being implemented. The future will be affected by retirement of old plants and construction of new ones for which different controls will be required. Generally, the additional emissions reductions will be more expensive, thereby increasing what an individual polluter would be willing to pay for additional allowances. Likewise, holders of ERCs are likely to increase the minimum price they are willing to accept as payment for these reduction credits. As a result, current holders of ERCs have some incentives to hold credits for appreciation. While this behavior is beneficial, in principle, in helping bring about the best allocation of emissions in response to intertemporal changes in regulations, regulatory uncertainty can lead to counterproductive permit hoarding.

Little can be done about the basic economic risks associated with the allowance market that are not caused by regulatory uncertainty. As in any asset market, allowances will fluctuate in price in response to changes in supply and demand factors. Participants in the allowance market can be left to themselves to determine how this market risk will translate into prices. Price fluctuations resulting from anticipation of future changes in control costs and other economic factors that will affect the value of the allowances over time are probably healthy and should not be discouraged.

More difficult to resolve are the regulatory uncertainties that create additional risk. How can market participants speculate on future government policy toward the allowances? How can market participants hedge on future changes in emissions reduction requirements that could send the market for allowances widely in either direction? How can market participants hedge on how individual state regulatory commissions will treat the allowances in exercising their rate-making powers? A rational response may be for present holders of ERCs to sit on their credits, letting them be used up in-house as regulations become more stringent in the future. This course could be the safest way to proceed in the choppy waters of regulatory uncertainty.

The Property Rights of the Emissions Credits May Be Degraded by Changes in Air Pollution Regulations as Part of Bringing About Further Reductions in Allowable Emissions, Serving to Reduce the Value of Emissions Credits

If the temptation is resisted to degrade emissions rights by declaring them to depreciate when more stringent controls are introduced, they should become more valuable as emitters seeks ways to further reduce emissions. However, if a current right to emit 100 pounds of pollution is declared to become a right to emit only 85 pounds when allowed emissions are reduced by 15%, the value is subject to degradation and possibly by much more than the increase in value because of increased demand for credits from more stringent regulation. The fall in value due to property right degradation is a full 15%, while the rise in value due to increased demand for credits is effectively the rise in marginal cost of controls, which provides the alternative to acquiring emissions credits, as emitters move to reduce levels of pollution. The degradation effect will outweight the increased stringency effect if the elasticity of marginal costs with respect to emissions reductions is less than unity, making costs rise by less than a full 1% when allowable emissions are reduced by 1%, as seems plausible.

What Happens to Emissions Credits if a Shutdown Occurs? Would the Credits Be Available for Trading or Sharing Across State Lines?

The decline in traditional manufacturing in the Chicago area is a fact of life that is likely to continue and to affect the demand for allowances. When an emitting plant closes down and leaves the Chicago area permanently, its emissions rights can be destroyed, given free to remaining emitters, perhaps on a proportional basis, or sold by the plant leaving the area to the highest bidders remaining in the area. The latter is most desirable from the point of view of efficient environmental control. A further question is whether the newly available emissions will be able to be used across state lines within the metropolitan area, which again seems desirable from an economic perspective but may raise political problems. Until some decision is made on these issues, there will be additional regulatory uncertainty that could inhibit participation in the allowance market.

Recessions Can Delay the Planned Use of Credits. Would Credits Be Forfeited in This Case?

When a plant leaves the area, it is clear that a permanent shutdown has occurred. The situation is less clear for a plant remaining in the area that may go for some time without using its emissions rights. Given the horizon of the emissions markets, which will extend into the next century, economic activity will oscillate over the future in ways that are not entirely predictable. A firm's output and, therefore, its emissions may be lower than originally anticipated. Purchase now of rights to emit in a future year could be a waste if output turns out to be lower than planned. The possibility of selling unused credits is fraught with uncertainty as to price as well as costs of arranging sales. Many smaller firms could be left in a relatively worse position during down periods than larger firms that are better equipped to diversify their resources away from the most negatively affected activities.

Additonal Modeling and Analysis of Polluter Behavior Are Needed

Formal simulation modeling of what can be hoped for from emissions trading, based on rational actor behavior and ignoring the impediments to participate we have emphasized above, overstate the gains from emissions trading. It would be feasible and desirable to include in formal simulation models explicit behavioral responses to the impediments, leading to more realistic estimates and better design of the mix of policies of which emissions trading will be a part.

The Advantages of Charging for the External Cost of Pollution

Recent state regulatory activity in the utility industry concerned with the environment has mandated the quantifying of external costs in the production of electricity, for example, the environmental costs imposed on others of burning a ton of coal connected with an extra cost of producing a kilowatt of electricity. This cost is to be added to the usual market or private cost of producing electricity so that the total cost, including environmental costs, can be used in planning fuel use in the production of electricity, taking account of environmental effects. An important development here is that, perhaps for the first time, dollar values attached to environmental effects are being used in a practical policy-making context. This state of affairs is helping to demonstrate the practicality of quantifying externalities in dollar terms, paving the way for the approach of charging for the external cost of pollution, long advocated by economists but heretofore dismissed as impractical.

The idea of a pollution charge or tax is venerable. The idea precedes emissions trading by 100 years and would actually reduce or eliminate the need for emissions trading. Charging for emissions can create incentives to reduce emissions to the point at which added reduction costs equal the charge, which tends to equalize marginal costs among emitters and allows flexibility in their responses. Combined with estimates of damages, this approach can aim to induce emitters to pollute up to the point at which the marginal damages caused by additional emissions equals the marginal control cost. In addition to reaching stationary sources, the approach can reach mobile sources through taxes on fuels and vehicles related to their environment damage.

Emissions charges have been the policy implication most widely discussed in the economics literature growing out of Pigou's distinction between private and social costs, which was the origin of environmental economics. This approach warrants serious continuing study to consider appropriate charge levels, monitoring, and benchmarking. Economically determined emissions charges are to be distinguished from punitive charges sometimes levied. More widespread use of emissions charges may have been resisted because of hesitancy to collect sums of money based on the effects of pollution which, at best, can be estimated only roughly. The sums of money may well be no greater than those paid by firms to meet mandated requirements to install pollution control equipment, which are used in place of emissions charges. A way to ensure that charges will be less is to make deviations in emissions from benchmark levels the basis for charges. The benchmark would be determined by a firm's property right to pollute, which could conceivably be its historical emissions but could be different. Firms would have incentives to find least-cost ways of controlling emissions and would either be taxed or rebated, depending on whether they exceeded or fell short of the benchmark level of emissions.

The Challenge of Convincing the Public that Markets Can Achieve Better Outcomes than Direct Controls

Emissions trading sounds attractive to those interested in economic analysis but, just as business people may be skeptical, so may the public at large. The problem of convincing the public may be underestimated and deserves further attention.

Conclusion

This paper has presented 14 points that, we believe, need to be considered in more detail in developing and improving emissions trading markets. Since it is not clear that all the problems can be overcome, it is also not clear that emissions trading can emerge as the panacea for controlling health risks from sulfur dioxide and particulates, acid rain, urban ozone, and other environmental effects.

Emissions trading is beginning to find a niche as one of the approaches to dealing with the environment but, realistically, traditional command-and-control methods of environmental regulation will probably also remain. Our most important point is that the role of emissions trading will be strengthened if the problems we have discussed are dealt with. Federal and state agencies are faced with the challenge of doing so. Help is needed from the academic community. A system of emissions charges or a hybrid system that combines emissions trading, emissions charges, and a minimum of command-and-control could be promising.

REFERENCES

Environmental Defense Fund. 1994. *Emissions Trading in Nonattainment Areas.* New York: Environmental Defense Fund.
Puller, Steven. 1993. "Valuing Environmental Externalities of Electricity Production in Illinois." Unpublished manuscript. University of Chicago.
Sholtz, A. M. 1994. "Observations for Environmental Markets: RECLAIM and Beyond." Unpublished manuscript. Division of Humanities and Social Sciences, California Institute of Technology, Pasadena.
South Coast Air Quality Management District. 1993. RECLAIM, Vol. III: Socioeconomic and Environmental Assessments, Draft Report. Diamond Bar, CA: South Coast Air Quality Management District.
South Coast Air Quality Management District. 1996. Annual RECLAIM Audit Report. Diamond Bar, CA: South Coast Air Quality Management District, p 199.
Tolley, G., J. Wentz, S. Hilton, and B. Edwards. 1993. "The Urban Ozone Abatement Problem." In *Cost Effective Control of Urban Smog,* edited by R. F. Kosobud, W. A. Testa, and D. A. Hanson. Chicago: Federal Reserve Bank of Chicago, pp 9–28.

2.8

❏

A DELIBERATIVE OPINION POLL
OF EXPERT VIEWS ON
ENVIRONMENTAL MARKETS

Richard F. Kosobud and Jennifer M. Zimmerman

❏ During the Workshop meetings, environmental markets attracted the most attention, resulted in the most spirited debates, and generated the most questions about future performance. Here were a group of well-informed participants representing all the concerned communities who had heard the arguments for and against markets from all sides and had deliberated over the important issues. It was too good an opportunity to miss. The editors decided to carry out a deliberate opinion poll on the problems and prospects of the three cap-and-trade markets for air quality control—for smog, acid rain, and climate change—that had occupied so much of the time.

Several features sharply distinguish this survey from the traditional design. It is not a survey of a random sample of the population, and its results cannot be projected by any sort of adjustment to the larger community. Rather, it may be considered a survey of an expert, though varied, group of Workshop participants who were exposed to and deeply involved in the exchange of ideas about environmental markets. Their responses, in our view, provide a significant new source of information on the strengths and limitations of environmental markets. Furthermore, their responses suggest how the general public may react after being more fully informed and given the opportunity for deliberation.

Of the over 100 participants who make up the Workshop membership, 43 elected to participate in the extensive questionnaire on environmental markets. In addition, seven outside experts closely involved in the design of markets also participated. Of those surveyed, 18% came from environmental or public interest groups, 22% from governmental units at all levels, 24% from businesses, and 36% from academia. Many of the contributors to this volume participated in the survey, which will give the reader an idea of the respondents' background.

The survey asked for the respondents' appraisal of the three markets described above. In many instances, questions called for scaled answers amenable to statistical analysis. Most questions also contained an open-ended feature so that more wide-ranging answers could be expressed. Both types of answers are summarized briefly in the results that follow. Guidance for the survey was obtained from the Survey Research Laboratory at the University of Illinois at Chicago. The protocols for the survey are available from the editors on request.

The most frequently cited advantage of environmental markets was that they could, in principle, meet pollution control goals more cost-effectively than other instruments (92% of the panel noted this static efficiency argument). Next cited in importance was the possible stimulation of future innovative, cost-saving control technologies (46% cited this dynamic efficiency point). The decentralized nature of the market and the potential saving in administrative burden were found by 42% to be separate and distinct advantages.

Respondents observed that markets can provide a flexible and reversible way of dealing with scientific and other uncertainties and a means to hedge against an uncertain future. Markets appear to be more acceptable than their incentive-based rival — corrective taxes — perhaps because they seem to afford traders more anonymity in engaging in market transactions and more autonomy in deciding whether to control or trade. One respondent mentioned that markets avoided some of the favoritism or special-interest biases that arise with traditional control measures, where waivers, manipulation, or grandfather clauses may be involved.

The most frequently mentioned disadvantage (34%) was the possibility of high private search and negotiation (transaction) costs. Also noted were questions about the fairness of initial allocations of permits: Several respondents thought that permits should be auctioned in all cases and not allocated free of charge. A few respondents were concerned that granting emitters the right to bank allowances over any time horizon could give rise to a burst or spike of emissions at the wrong time. However, most believed that the banking of permits should always be allowed, acting, as it does, to remove the pollutant faster. The adequacy of measuring, monitoring, and enforcement procedures also emerged as matters of concern. Regulatory uncertainties in the utility industry were listed as bearing adversely on the performance of the sulfur dioxides (SO_2) market, thus raising the important problem of the compatible adaptation of the new markets to the existing layer of conventional regulation.

PUTTING THE CORRECT CAP ON A CAP-AND-TRADE MARKET

Establishing an aggregate cap on pollution requires a judgment about the relationship of the benefits of reducing damages to the costs of control. The Clean Air Act Amendments of 1990 called for a roughly 50% reduction of national SO_2 emissions and about a 40% reduction of urban ozone (O_3) concentrations in the Chicago severe nonattainment region. The Administration's "Climate Change Action Plan" (Clinton and Gore 1993) called for a stabilization of 1990 U.S. greenhouse gas emissions by the year 2000. What do these caps imply about the amount of control resources that will be applied to these problems compared with the benefits of pollution reduction obtained? Will the cap overcontrol, undercontrol, or be about right?

In response to a question about the appropriate degree of control in the three markets, the answers, summarized in Table 2.8.1, reveal that the majority of respondents who had an answer thought that the degree of control in these markets was about right. Only in the case of global warming policy was there significant concern (by 21% of the panel) that overcontrol was, in fact, the case and would consume resources better used elsewhere.

Highly significant to the editors is the large number of respondents who did not have an option on this matter, which is so critical for rational allocation of resources. Is this evidence of the lack of information (and required research) on the impacts of pollution and the benefits of its reduction, even among informed observers?

One of the lessons of Table 2.8.1 may well be that cap-and-trade markets will encourage clarity in thought about the permissible aggregate level of emissions or concentrations of a pollutant, the point at which benefits and costs are matched. Conventional regulation, oriented as it is toward microstandards or technologies that too often focus on the rate and not the total volume of emissions, may stand in the way of this important general evaluation.

Table 2.8.1. Are policy targets (caps) for pollution reduction set too high, too low, or about right? (in percent of respondents)

Degree of control	SO_2 market (%)	O_3 market (%)	CO_2 market (%)
Too low (undercontrol)	0	9	9
About right	35	26	26
Too much (overcontrol)	9	9	21
Don't know	56	56	44

Note: Number of respondents = 50

VIEWS ON THE SO_2 MARKET

The respondents were asked if they were surprised by the prices revealed in the first auctions of the set-aside sulfur dioxide allowances (2.8% of the total allocated) and by the number of private offerings sold at these auctions. The purpose of these auctions, which were managed by the Chicago Board of Trade for the U.S. EPA, was to facilitate price discovery in the market and to provide another entry channel for newcomers to the electricity market, such as independent power producers. Net revenues from auctions were to be returned to utilities in proportion to their contribution of auctionable permits.

The panel, in general, was not surprised by these revealed prices; only 24% were surprised by the apparently low prices bid, and 22% reported no opinion. A number of reasons were noted for these prices and for the few private offerings traded in the first four auctions. The relatively small cutbacks of emissions in the early stages of the program, which may have prompted companies to seek technological solutions instead of using the market; shyness in bidding (public questioning of utility trading); permit hoarding against uncertainty; the behavior of utility regulators; and declines in coal and scrubber prices were cited by participants. On balance, the common view was that the market was achieving significant savings.

Respondents were asked whether utility regulation—state and Federal Energy Regulatory Agency (FERC) decisions—was affecting the market. The great majority, 85%, viewed these decisions or lack of them as having an adverse effect, limiting the welfare gains of the market. Reasons given for this effect were varied. Almost a third of the respondents believed that the present lack of appropriate regulatory decisions made utilities conservative—hesitant to sell or buy. Regulatory decisions on the proper accounting procedures for transactions and on the allocation of gains or losses to ratepayers or stockholders remain to be made or are not yet clear. Variations among state regulatory decisions could hamper interstate trading.

Monitoring of emissions, enforcement of market rules, and establishment of penalties for a source emitting more than allowed were serious matters for Workshop consideration. The U.S. EPA has promoted a continuous monitoring emission technology for the larger emitters. The penalty for overemitting has been set at $2,000 per excess ton, with the proviso that the excess be made up by an identical reduction in the next period. Panel respondents were generally of the view that emissions could be measured accurately enough for control purposes (only 6% had doubts on this score) and that the fines were appropriate (only 9% thought them too lenient). These views are itemized in Table 2.8.2. However, there were concerns expressed about the lack of information about the amount of bankable allowances being held by specific utilities.

There was strong support for the development of futures and forward markets for permits: 56% of the panel thought such derivatives to be a possible and desirable development; 12% were of the maybe yes–maybe no view; and only 18% believed such development was not possible. Only 15% did not have an opinion on this

Table 2.8.2. Views on the SO_2 market (in percent of respondents)

Enforcement (%)		Monitoring (%)		Futures and option (%)	
Too lenient	8	Very accurate	24	Very possible	34
Somewhat lenient	2	Somewhat accurate	30	Somewhat possible	22
About right	54	In between	22	In between	8
Somewhat strict	2	Somewhat inaccurate	4	Not very possible	20
Too strict	2	Very inaccurate	0	Not at all possible	0
Don't know	32	Don't know	20	Don't know	16

Note: Number of respondents = 50.

seemingly esoteric point. One respondent noted that futures markets, and forward markets for delivery, seemed feasible but doubted that the number of transactions and price volatility would justify an options market. These derivatives, if developed, provide opportunities for improved risk management, an advantage of environmental markets clearly perceived by the panel and not possessed by other incentive approaches.

What confidence in the SO_2 market did respondents have after two years of deliberation and review of the early evidence? Fifty percent have had their confidence strengthened as against 18% who have had their confidence weakened; 24% had their confidence unchanged. Only 8% reported "Don't know." This is important evidence on the change in confidence brought about by more information on, and deliberation about, environmental markets. Is it reasonable to conclude that more public experience with this market could bring even more support?

VIEWS ON THE URBAN O_3 PRECURSOR MARKETS

The Workshop conference in June of 1993 presented Mary Gade, director of the Illinois Environmental Protection Agency (IEPA), and Sam Skinner, CEO of the Commonwealth Edison Company (ComEd), who jointly announced a new program to introduce tradable permits for nitrogen oxides (NO_X) control among stationary sources in the Chicago region. Eighteen months later, a revised plan was announced to introduce tradable permits for VOM or VOC (volatile organic materials or compounds) control among stationary sources. The change was due largely to new scientific findings from the Lake Michigan Ozone Study (LMOS) that were made available in the interval. That study was based on emissions projections and an advanced chemical, physical, and meteorological modeling of the region's ozone formation processes.

Unexpected new scientific findings were promptly translated by the IEPA into a revised environmental market. A majority of the panel (60%) agreed that the findings of the LMOS study should be, and were properly regarded as, essential components in redesigning the ozone control market. The participants mentioned the spatial findings of the study, that unilateral NO_X emission (and, hence, concentration) reductions in particular areas would not decrease ozone concentrations and could actually increase ozone in certain areas. A fifth of the panel reported that the boundary conditions (the

Table 2.8.3. Views on the O_3 Market (in percent of respondents)

Return to traditional regulation		Combine with vehicle fleet controls		Combine with "cash for for clunkers" incentives	
Agree	10	Definite yes	34	Definite yes	28
Partially agree	8	Qualified yes	22	Qualified yes	14
In between	22	In between	14	In between	12
Partially disagree	28	Qualified no	6	Qualified no	12
Disagree	28	Definite no	2	Definite no	14
Don't know	4	Don't know	22	Don't know	20

Note: Number of respondents = 50.

finding that concentrations of precursors and ozone coming into the nonattainment area were very high) were important, suggesting that interstate or out-of-state control measures and trading were also required for control.

By no means would a well-functioning market for VOM emissions from stationary sources alone, perhaps 30% of total VOM emissions, bring the region to attainment. The panel's interest in the simultaneous control of VOM and NO_X, in interpollutant trading, in giving credit to stationary sources' efforts to reduce mobile source emissions, and in control of emissions outside the region attest to the Workshop's probing of the complexity of cost-effective control of ozone and to the work ahead in implementing this environmental market.

Table 2.8.3 provides a summary of the panel's views on whether the new scientific findings point toward a return to conventional regulation (18% were of this view) and whether combining markets with other forms of controlling mobile source emissions was feasible.

VIEWS ON THE CO_2 MARKET

In some respects, anthropogenic climate change meets many of the tests for application of market-based approaches, and for cap-and-trade markets in particular, even though it is an uncertain and very long-run problem. There are numerous greenhouse gas sources, especially for CO_2, with widely varying control cost relationships. Fossil fuels, the source of the leading greenhouse gas, are well documented. Greenhouse gases disperse rapidly and uniformly in the upper atmosphere; there are no "hot-spot" problems requiring differential control of greenhouse gas emissions in particular areas although regional damages or impacts may differ markedly.

The uncertainties at every level of knowledge of this global issue imply a dynamic milieu for policy decisions favoring flexibility— attractive features of market-based approaches. Risk management may be furthered by developing futures and forward markets based on the cash-tradable permit. Market-based approaches may also be more readily accepted in this area, where other regulations would be resisted. These considerations help explain the growing interest in incentive schemes to deal with global warming.

Furthermore, the 1992 Rio Summit's Framework Convention on Climate Change, in providing for "joint implementation," laid a partial groundwork for constructing a market in country emissions quotas. The main idea is to enable developed countries and associated enterprises to earn emissions reduction credits by furnishing, at favorable prices, technology and capital to other countries, enabling both countries to share in the reduction of greenhouse gas emissions. Savings are realized by the more numerous opportunities to achieve cost-effective reductions in emissions in the less developed or formerly centrally planned economies.

The ground rules for the U.S. Initiative on Joint Implementation require eligible U.S. participants in their negotiations with (developing) countries to describe specific measures for reducing or sequestering emissions, to provide data on baseline and future emissions both with and without control, and to establish provisions for tracking and verifying emissions. However, the granting of tradable "credit" for any transactions under the initiative is yet to be clearly established. The market, at best, should be classified as nascent. It is significant, therefore, that 56% of the Workshop respondents felt that joint implementation was a promising beginning to a trading scheme for controlling global warming although 30% had reservations. Several respondents mentioned projects already under way between U.S. enterprises and counterparts in Mexico, Russia, Costa Rica, and the Czech Republic.

Juxtaposed with these positive views are the comments of respondents about the difficulties that could lie ahead in CO_2 trading programs: resolving the deep uncertainties of the problem area sufficiently to secure support for action; negotiating a binding international agreement; collecting and monitoring emissions data; and enforcing market rules on the international scene.

WILL CAP-AND-TRADE MARKETS PROVE TO BE A SUCCESSFUL INNOVATION?

The panel was asked to estimate the chances of success for each of the three markets, given what they know now. Success was defined as an "efficient" market within which the environmental quality goal is achieved, prices reveal control costs, and trades enable emitters to reduce their costs; consequently, welfare gains are larger than those obtainable with alternative approaches. Chances were to be scaled from 0 to 10 (from no chance to certainty).

Figure 2.8.1 gives a three-dimensional distribution of the chances. The volume of points rises toward the higher values, indicating a favorable forecast for the markets in general, but there are healthy variances and significant differences among the markets, suggestive of the wide range of concerns about this new policy tool. The mean chances are highest for SO_2 at 6.6, next for O_3 at 5.2, and lowest for CO_2 at 3.4.

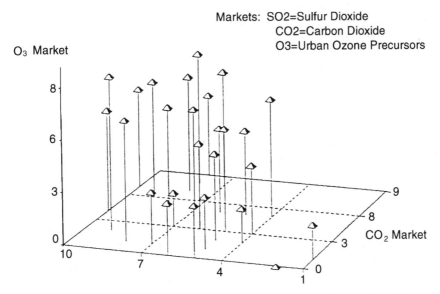

Figure 2.8.1. Chances for market success: Respondents' estimates of market success on a scale from 0 to 10. Number of respondents answering equals 36. Two or more respondents giving the same probabilities are not separately recorded.

SIGNIFICANT INTERRELATIONSHIPS AMONG
RESPONSES

Is there a pattern among the answers that would suggest coherence or consistency in the views expressed? We explore this question in a statistical sense, using a multiple regression tool of the single-equation least-squares variety. A natural dependent variable is the respondent's estimate of the chance of success of the market. Several caveats are in order. Some important questions led to answers that defied numerical scaling. This was especially true of the answers to questions about the government regulation of utilities in the SO_2 market and about the importance of new scientific findings in the O_3 market. These obvious candidates for the role of significant determinants of success estimates had to be tested by reading and reporting the answers qualitatively, as we have done.

Furthermore, numerical scaling is, to a considerable extent, arbitrary. The dependent variable, the chance of success, is a limited dependent variable that ordinarily calls for use of specialized statistical tools. On inspection of the distributions of the answers, and after some experimentation with alternative estimation methods, the editors decided that it was safe to report the results of the simpler estimation method. Explanatory variables, where it appeared sensible, were scaled 1 through 5, or as dummy variables. Interested readers may request data from the editors. Extreme caution is advised in interpreting the regression coefficients.

Finally, the statistical model assumes that what is on the right-hand side of the equation determines what is on the left; that seems logical to us, but it may be that all are jointly dependent, so that the left-hand variable, the chance of success as formulated in the respondent's view, determines responses to right-hand variables.

We present standard errors of estimate to test the significance of regression coefficients not on the basis that our panel is a sample from the general population but that the set of answers entered on the questionnaire was a drawing from nature's urn of all possible answers.

Despite all these reservations, we find the relationships uncovered to be of great interest. They indicate, as revealed in Table 2.8.4, that those respondents who believed that markets were cost-effective alternatives to other control measures gave significantly higher chances of success, as if this advantage would enable the market to overcome the inevitable initial obstacles. Taking these coefficients at face value, we note that a respondent's mention of

Table 2.8.4. Regression results: Determinants of respondents' estimates of chances of success

	Explanatory variable	SO_2 market	O_3 market	CO_2 market
1.	Markets are cost-effective	3.982 (.716)	8.302 (.864)	5.533 (1.102)
2.	Allowances as limited privacy property	−.867 (.200)		
3.	Utility regulation factors	+1.016 (1.229)		
4.	Confidence in markets strengthened because or SO_2 facts		+1.388 (.325)	+.193 (.487)
5.	Joint implementation is a promising start			+.684 (.324)
	Adj R-squared	.943	.899	.659
	Number of respondents	25.	29.	34.

Note on coding conventions:

1. Markets are cost-effective: coded 1 if mentioned, 0 if not.
2. Permits are not strictly private property: coded 5 if "very significant" through 1 if "very insignificant."
3. Utility regulation factors: coded 0 if factors were mentioned that would adversely affect the market, 1 if not mentioned.
4. Confidence strengthened because of SO_2 facts: coded 5 if "definite yes" through 1 if "definite no."
5. Joint implementation is a promising start: coded 5 if "very promising" through 1 if "not promising at all."

   Dependent variable is the respondent's estimate of the chances of success: coded 0 through 10.

   Intercept term suppressed; all "don't knows" excluded. Numbers in parentheses are standard errors.

cost-effectiveness on the average increased the chances of success by 4 points for the SO_2 market, by over 8 points for the O_3 market, and by 5.5 points for the CO_2 market.

On the other hand, those respondents who were concerned about the limited private property character of the SO_2 allowance gave lower chances of success to that market (the coding gave a 5 if the lack of private property character was viewed as very significant and ranged through 1 if not). While regulatory uncertainties were found to be of great importance in the qualitative analysis of the survey, the coding was such that no test should be devised. However, a crude dummy introduced to indicate regulatory uncertainty did produce a coefficient with the right sign, albeit not significant at the conventional level.

Note that those respondents who found their confidence in environmental markets increased by the performance of the SO_2 case, the most advanced market, assigned higher chances of success to the other markets. We interpret this as an indication of the importance of having successful examples if environmental markets are to be extended to other areas. Finally, those who found joint implementation ideas to be promising assigned higher chances of success to an ultimate CO_2 market.

REFERENCES

Clinton, William J., and Albert Gore Jr. 1993. "The Climate Change Action Plan." White House release. Washington, DC. October.

3

❏

REGULATORY INNOVATIONS

AWAITING TRIAL

❏

3.1

❏

INTRODUCTION TO PART 3: NEW PROBLEMS AND MORE ALTERNATIVES

The Editors

❏ The planned agenda of the Workshop was revised several times as new items drew increased attention, among them a growing, massive, but uncertain environmental problem—climate change; a perhaps underappreciated market-based tool—pollution taxes; and a regulatory reform not directly related to market-based approaches— voluntary pollution reduction. It was neither possible nor advisable to ignore these items. If market-created incentives are truly an advance, they should hold out the potential for application to new problems. If pollution taxes can work effectively in certain situations, they deserve more consideration. If incentives can be developed that encourage voluntary control efforts, they ought to be tested and tried. The latter could become complements or substitutes to market-based approaches.

The Workshop devoted a number of meetings to this extended list of topics. Although some may not appear to fit neatly into the framework of this book, the editors have decided that they can throw important light either on problems amenable to market-based incentives or on the utility of decentralizing control decisions. The studies selected deal with market-based designs to control climate change, with the use of green taxes as a source of revenues and a means for pollution reduction, and with a seminal pioneering effort to study ways to reduce and prevent pollution through more voluntary efforts.

213

ANTHROPOGENIC CLIMATE CHANGE
AND EMISSIONS TRADING

It may seem premature, at best, to devise incentive plans for control of induced climate change. The extent and timing of damages are highly uncertain, relatively unexplored, and even controversial subjects. Furthermore, it must be kept in mind that a limited increase in CO_2 concentrations would benefit some crops and that complete removal of greenhouse gases would render the earth uninhabitable. These gases are not like toxic pollutants in which zero presence would be biologically preferable; there are desirable concentrations of greenhouse gases to have present in the atmosphere. Problems arise, however, when those concentrations are exceeded. The timing of impacts, many of them perhaps 50 to 100 years ahead, is extremely long-run as environmental problems go. The "pollutants," largely carbon dioxide (CO_2) from fossil fuel combustion and other greenhouse gases generated by human activity, are rapidly dispersed in the global atmosphere so that no one country can do much by itself to contain the problem. If controls are to be undertaken, they will require reductions in the use of vital materials that are sources of CO_2 emissions, like fossil fuels. Securing international agreement on control measures and the amount of reduction will be difficult, at best.

No matter the uncertainties and difficulties, the continuing accumulation of greenhouse gases will not go away. Every measure reveals that these concentrations are increasing every year; there is little disagreement on this count. Many scientists and observers believe that the concentrations are approaching the point at which chances of long-run damages from global warming and other climate changes become very serious. The International Panel on Climate Change (IPCC), a group of prominent scientists set up to advise nations signing the Framework Convention on Climate Change, have issued background studies and, lately, a finding on the human-induced cause of the problem (Houghton et al. 1992, Weiss 1996).

Several economists have developed complex intertemporal economic-meteorologic models that can be used to generate simulations of the benefits and costs of alternate control policies. On the basis of these simulations, these economists have recommended starting with modest controls, such as low initial carbon taxes, which can then be tightened over time to achieve cost-effective control as needed (Nordhaus 1992, Cline 1992).

Other modeling endeavors have explored the use of tradable cumulative or evaporative emissions permits that are initially issued in an amount that would set a limit on the ultimate atmospheric concentration of a greenhouse gas such as CO_2. The climate change market becomes a variant of the cap-and-trade model. The permits are issued for emissions denominated in tons of carbon, which assumes, as is reasonably accurate, that the fraction of emissions that remains in the atmosphere is stable. The price of these permits should start low, given the large initial allocation, and rise as permits are used up by enterprises digging the coal, pumping the oil, or transporting the gas. The increase in price could be expected to be in line with the long-term interest rate as traders of permits calculate the advantage of either holding the depleting stock of permits and reducing emissions or turning in permits for emissions (Kosobud et al. 1994). This intertemporal price path of the gradually declining stock of permits could provide the energy sectors time to adjust to the increasing marginal cost of reducing greenhouse gas emissions.

The study selected for this volume develops a design for the mechanics of a cap-and-trade market approach to climate change management but does not specify possible limits to emissions or concentrations and their rate of change. These are left for negotiation. Drafted by Richard Sandor, a pioneer in financial commodity innovation for much of his career, the plan for CO_2 emissions trading could serve as one practical proposal for national and international negotiation on the use of markets. While he concentrates on CO_2 in this effort, Sandor is well aware of the importance of other greenhouse gases, such as methane and nitrous oxide. The market could be extended to other greenhouse gases once research under way has clearly established the relative warming potential of each gas. Additional research may indicate, in fact, that control measures other than markets will be more suitable for some of the greenhouse gases.

Tradable permits could help achieve cost-effective control both within a country and among countries as control costs clearly vary dramatically by industry, by sector, by region, and by country. The discussants of Sandor's study add their views on implementation concerns. There will be winners and losers resulting from climate change in view of the projected range of regional impacts. Initial allocations of tradable emissions permits could, in fact, be transfers of wealth among nations designed to secure cooperation, but their amounts might have to be very large. Having practical designs such as Sandor's on the table could be of great help in the long negotiation process that may lie ahead.

Another step in the process toward an effective international climate change control policy, if such proves both desirable and achievable, is considered in the paper by Edward Helme and Janet-Anne Gille. They discuss a project arising from the 1992 Rio Summit's idea of joint implementation, which may, in fact, be an early form of emissions trading. This is the Decin project involving a city in the Czech Republic and three midwestern utilities, the former gaining a low-cost loan for a cogeneration facility and the latter gaining possible carbon CO_2 emission reduction credits.

The authors take us through the negotiation process by which city and utility officials, with help from the Center for Clean Air Policy, determined the change from coal to natural gas fuel that led to lower CO_2 emissions per unit of energy input. The authors indicate the kinds of costs and capital resources required for the change. The availability of external financing from the utilities, at terms more favorable than otherwise obtainable, enabled the city of Decin to benefit from a project that reduced many of the pollutants in the air, in addition to CO_2, that had long harmed the residents.

Helme and Gille believe that the variation in control costs of greenhouse gas emissions around the globe creates strong incentives for further development of joint implementation projects. When enterprises and countries can achieve cost-effective reductions of greenhouse gases by securing technical help, technology, and capital from more advanced enterprises and countries, which can then share in the credits from such greenhouse gas reductions, a potentially tradable credit is, in effect, created. The authors are careful to point out the economic and political difficulties that must be overcome before such projects become a significant factor in climate change control.

POLLUTION TAXES

Using environmental markets to succeed when ordinary markets fail due to pollution externalities works by correcting prices as discussed in the introduction to Part 1. The same effect can be achieved, in principle, by levying the correct tax on the pollutant. The tax rate should be set so that the estimated marginal harm caused by the last unit of pollutant reduced by the tax just equals the marginal cost of that reduction. This would also be the price of the tradable pollutant permit if it is issued in volume that constrains the pollutant to the same degree as the tax rate. The prices, which were wrong for environmental quality before use of tax or tradable permit scheme, become right after the instruments are applied.

We also discussed, in the introduction to Part 1, the pros and cons of taxes versus permits as instruments of control. The best choice, from the economic point of view, would appear to depend on the particular situation, the nature of the pollutant, the shapes of the cost and harm functions,[1] and the nature of the industry and enterprise. It is arguable that corrective taxes, suffering from the unpopularity of taxes in general, are less frequently in the environmental arena than their merits warrant. Note that we distinguish corrective taxes from the frequently used levies and fees that generate administrative revenues but are not sufficient to remove externalities. Unlike environmental markets, whose effects on commodity prices are not readily discernible, corrective taxes have highly visible effects on the taxpayer and, frequently, on prices.

Robert Repetto develops the idea of green taxes as a means not only of correcting prices by internalizing the social costs of the pollutant but also of providing a revenue stream that can replace other taxes that distort incentives, for example, to work or save. That is, taxes on pollutants distort prices but in the right direction by reducing pollution compared with taxes on savings or work effort that distort in the wrong direction. That green taxes generate a revenue stream that can substitute for taxes in other areas may make them more acceptable to the general public if knowledge of this trade-off becomes widespread and credible.

Repetto marshals as strong a case as he can for green taxes, citing evidence for the distorting effects of existing taxes and the potential revenues from pollution taxes. One discussant indicates that he would prefer a lower estimate of the marginal excess burden of labor and capital taxes than that used by the author. Repetto provides an extensive list of references that includes economists who do not agree with him on this and other matters. He thus affords the reader access to the current range of opinion on the properties of green taxes while laying out the case for their increased application.

[1]Weitzman (1974) has an interesting analysis of this point.

VOLUNTARY POLLUTION REDUCTION

Turning to voluntary efforts to improve environmental quality in place of conventional regulation, Ronald Schmitt provides an updated account of the path-breaking Yorktown project. One of the first comprehensive projects of this nature, it appears to have opened the door to decentralized and cooperative approaches to pollution control and prevention that go well beyond ordinary command-and-control (CAC) regulation.

The Amoco Oil Company/U.S. EPA collaboration on a study of pollution control at a refinery presents facility-type information for answers to two fundamental questions of environmental regulation. First, what has existing command-and-control regulation of the refinery actually accomplished in terms of reducing pollution and in terms of furthering our understanding of the costs of specifying particular control practices and technologies? Second, if a fresh start were made in gathering data on all pollutants, be they emissions in the air, effluents in the water, or contaminants in the soil, and if the most efficient control practices and techniques were applied to control of these pollutants, would the resultant pollutant reductions and costs of control resemble the present CAC scene?

Most observers from all sides have concluded from the project reports that the present CAC directives leave much to be desired. Some of the most significant sources of emissions to the air were not subject to regulation, and a number of sources, frequently insignificant, required cost-ineffective measures. The U.S. EPA, with the best of intentions, had neither the knowledge nor the resources to duplicate the detailed management of the refinery that would be necessary to control pollution to the degree required. The Amoco Company had neither the incentive to go beyond existing regulations nor the desire to make innovations under CAC that might well result in tightened requirements.

As Schmitt explains, the Yorktown project objective was to obtain information relevant to these questions and not explicitly to revise current regulation. To make quick regulatory changes on the basis of Yorktown would have raised serious questions of precedent and regulatory design for the U.S. EPA. To delay making quick use of the knowledge gained proved frustrating to the emitter. The benefits of the experiments, consequently, were not so much in revising the controls on the Yorktown refinery but in serving as an inspiration for a series of later initiatives among emitters and governments to work out voluntary and cooperative agreements for the improvement of environmental quality that replace or go beyond centralized regulation.

Examples of efforts that can be related to Yorktown may be found in the pollution prevention program of the Great Printers Project (1994) and in the administrative initiative provided in the U.S. EPA Excellence in Leadership (XL) program (1995). The incentive for emitters is that, by such voluntary efforts, they gain in flexibility of environmental management and in the potential for realizing savings. These are advantages that resemble the incentives created by market-based approaches. Furthermore, voluntary efforts can frequently be dovetailed neatly into market-based approaches (credits can be given for voluntary efforts that reduce emissions below performance standards).

Ronald Schmitt, by bringing the Yorktown project up-to-date, offers guidance for those interested in building further on this foundation. Appropriately, several representatives of both the regulating and public interest communities involved in the Yorktown project act as discussants to contribute their views on the implications of this pacesetting experiment. Governments gain in the case of successful voluntary efforts by an easing of the unpopular regulatory burden while retaining, as is essential, the goal-setting, monitoring, and enforcement roles. Through cooperation of the concerned communities, savings beyond conventional regulation would appear to be possible. Many questions remain, however, on the long-run effectiveness of such endeavors, on the magnitude of savings realized, and on the costs. No one is yet quite sure how to work these efforts into the existing regulatory fabric. These issues will have to be resolved before voluntary measures find their place among other regulatory tools.

REFERENCES

Cline, William R. *The Economics of Global Warming.* 1992. Washington, DC: Institute for International Economics.

Council of Great Lakes Governors, Printing Industries of America, and the Environmental Defense Fund. 1994. *The Great Printers Project.* Minneapolis, MN: General Litho Services. July.

Houghton, J. T., B. A. Callendar, and S. K. Varney, ed. 1992. *Climate Change 1992,* The Supplementary Report to the IPCC Scientific Assessment, Intergovernmental Panel on Climate Change. Cambridge, U.K.: Cambridge University Press.

Kosobud, R. F., T. A. Daly, D. W. South, and K. G. Quinn. 1994. "Tradable Cumulative CO_2 Permits and Global Warming Control." *The Energy Journal* 15(2) 213–232.

Nordhaus, William D. 1992. "The Dice Model: Background and Structure of a Dynamic Integrated Climate Economy Model of the Economics of Global Warming." New Haven: Yale University, mimeographed. February.

U.S. EPA (Environmental Protection Agency). 1995. "Regulatory Reinvention: Excellence in Leadership (XL) Pilot Projects." *Federal Register*, May 23.

Weitzman, Martin L. (1974). "Prices Versus Quantities." *Review of Economic Studies* XLI. October. 477–91.

Weiss, Peter. 1996. "Industry Group Assails Climate Chapter." *Science* 272(June 21):1734.

3.2

❑

TOWARD AN INTERNATIONAL CO₂ ENTITLEMENT SPOT AND FUTURES MARKET*

Richard L. Sandor

❑ The development of new types of commodities and markets emerges as a consequence of significant economic and/or political structural change. A study of the history of organized markets, ranging from futures and options on the Dutch East India company in Amsterdam in the seventeenth century to rice futures in Osaka in the eighteenth century to the emergence of modern futures trading in wheat in Chicago during the nineteenth century, illustrates these structural changes. As a result of high levels of economic activity in the twentieth century, nations around the world have become more aware of the need for sustainable development. These are the same kinds of structural, economic, and political changes that have caused new markets to emerge. Our attention now is drawn to the fact that air and water are simply no longer the "free goods" that economists once assumed. They should be defined as "property rights" so that we can develop a new market that will lead to a least-cost solution to global warming.

Economists and practitioners have been increasingly concerned with how to reduce global warming costs effectively. This paper assumes that this reduction can be achieved through the creation of a tradable entitlement market for carbon dioxide (CO_2), a leading

*This chapter draws on prior publications (Sandor 1992a, chap. 9) and (Sandor 1992b, pp. 23, 24).

global warming or greenhouse gas. Our purpose here is to begin a discussion of the mechanics of such a market. It is by engaging in this dialogue that we may begin to address the issue of why constructing environmental markets has not figured prominently in the excellent literature of the economics profession with regard to externalities. The simple answer is that there is no theoretical construct in economics to establish markets (Coase 1992). However, custom and practice in the futures industry provide us with the insight necessary to establish new markets.

The first step in developing an entitlement market in CO_2 is to define the commodity. Once the commodity has been defined, we can turn our attention to implementing market architecture for spot and futures markets.

It is important to emphasize that, over the course of the twentieth century, the definition of a commodity from which a spot market can evolve into a futures market has changed dramatically. By the middle of the twentieth century according to Blau (1944–1945), "the range of commodities eligible for futures trading is limited to goods in general demand for which a world price level can be established and which are capable of a high degree of standardization in quantity and quality. These conditions only apply to certain categories of foodstuffs, raw materials and highly standardized semimanufactures entering international trade." Blau's definition had been extended by others so that the general consensus at midcentury appears to be that the definition of those commodities that lent themselves to successful futures markets was confined to primary bulk (agricultural or metal) or semiprocessed storable commodities. A wave of successful innovation during the second half of the twentieth century has expanded limits implied by this definition. The introduction of a successful plywood futures contract eliminated the restriction that a commodity must be, by definition, primary or semiprocessed in character. The success of cattle and hogs eliminated storability. Active trading in foreign currencies struck bulk from the definition of a tradable commodity. Finally, with the introduction in 1975 of futures markets in Ginnie Maes, government-insured mortgages, the very word *commodity* became invalid.

In recent years, we have witnessed another watershed in the definition of a commodity. The Chicago Board of Trade has received approval from the Commodity Futures Trading Commission for a futures market in the tradable allowances of sulfur dioxide, a cause of acid rain. This property right was created by the United States Congress through the passage of Title IV of the Clean Air Act Amendments of 1990 (CAAA).

This represented the first attempt to establish organized markets in the trading of emission allowances resulting from assigned property rights. Those who might draft a global warming treaty should carefully study the Clean Air Act Amendments in their entirety. They can effectively be used as a prototype for defining a commodity and can therefore provide the framework for creating a tradable carbon dioxide spot and futures market, carbon dioxide being a major greenhouse gas, amenable to emissions trading control.

Since carbon dioxide emissions occur globally, the first step in accomplishing the creation of a tradable carbon dioxide entitlement would be to draft a global warming treaty that addresses applicable concerns and delineates the basis for the creation of the "property right," or commodity. Ideally, the treaty should also provide for the creation of a United Nations agency to govern and enforce the provisions of the treaty, possibly to be called the United Nations Global Environmental Protection Agency (UNGEPA). In January of 1995, a workshop on Economic Instruments for Sustainable Development (organized by the Czech Government under the auspices of the Commission for Sustainable Development) endorsed proposals for a pilot program for trading CO_2 entitlements, including the formation of UNGEPA.

The process must begin by determining who can participate in the market. Participation in the pilot program should be on a voluntary, nonbinding basis. The UNGEPA could assign a fixed amount of carbon dioxide property rights, or "allowances" (sulfur dioxide tradable entitlements are called allowances under the CAAA) to all signatories of the pilot program. It could also provide for setting aside a fraction of those allowances to prospective signatories. The analogy in the CAAA is the "opt-in" provisions that entitle additional companies and industries to participate. The main findings of the Prague Workshop also encouraged governments to facilitate and promote participation.

Additionally, a baseline must be established to define the measure of global warming. The granting of allowances should be based on existing and/or future acceptable levels of national and global emissions. The "baseline" level of carbon dioxide emissions should be estimated from a prior year, for example, 1990, based on factors such as fuel type and consumption, specific technology in use and design, capacity utilization, age of technology, and the like. Implementation should be planned for a future date, such as 2000.

The equitable allocation of carbon dioxide emission allowances, initially and over time, will be important in terms of perceived fairness and the political acceptability of the control mechanism.

The UNGEPA could establish and administer this distribution task. From a trading perspective, initial allocations need be constrained only by the necessity of establishing a competitive market. However, consistency in the allocation process over time will be essential for the determination of value. Uncertainty and incongruity in the supply of allowances will serve to hinder the exchange of allowances between private parties in both the spot and futures markets. The issue of initial allocations will require intensive bargaining since it is important that less developed countries do not perceive that their quest for industrialization has been hindered. Since the initial patterns will implicitly provide a basis for nonsignatories ultimately to become part of the process, the allocations should be quantitatively based. For example, projected future emissions based on productivity, population, and density of developed nations could provide guidelines.

U.S. legislation provides a historical precedent for establishing a baseline. The Clean Air Act Amendments set as a primary goal the reduction of annual national sulfur dioxide emissions by 10 million tons below the level that was emitted in the mid-1980s. To achieve these reductions, the law requires a two-phase tightening of the restrictions placed on fossil fuel–fired power plants. Phase I began in 1995 and Phase II begins in the year 2000. Utilities are fined if their emissions exceed enumerated tonnage levels. To exceed enumerated tonnage without penalties, they would need to purchase emission allowances from units with excess allowances. The two-phase features of the CAAA was implemented in order immediately to reduce emissions from the highest-emitting plants and to develop experience for the participants. Careful examination of whether a carbon program should follow this procedure is a prerequisite of any global treaty.

Standardization is critical and implicit in all aspects of the creation of a global carbon dioxide commodity market, from the method and formula for monitoring emissions to the issuance of the tradable entitlements. This suggests that the tradable entitlement be issued by a single entity, in this case, the UNGEPA; otherwise, segmentation might result from variation in issuance standards and credit standing among participants, and so forth.

The initial term of the pilot program should be 15 years. Annual permits should be issued. Annual permits issued with a program life of 15 years provide the flexibility to deal with issues of development and increasing knowledge of the science and economics of global warming.[1] Certainty of 15 years of property rights will permit market

---

[1]For an analysis of alternative criteria, see Grubb and Sebenus (1992).

participants to make both investment and trading decisions. The 15-year term also seems reasonable in light of the extrapolation of value based on the yield curve for government debt. From an operational viewpoint, entitlements can be established at any phase in the life cycle of carbon dioxide gas; extraction or production, consumption or emission. Emission entitlements do appear to offer the least-cost means of control since they are relatively homogenous and readily monitorable at any stage, as discussed in Tietenberg (1993). This scheme provides for a leveraging of the economic infrastructure that is in place for implementing the CAAA of 1990. Liquidity would be enhanced by spread trading between other emissions entitlements, such as sulfur dioxide or other greenhouse gas markets, as they emerge. For example, an Arizona utility bartered CO_2 allowances with a New York utility in exchange for SO_2 allowances.

IMPLEMENTATION OF SPOT MARKETS

For purposes of this chapter, spot markets will be defined as standardized contracts for the immediate delivery, in exchange for cash, of allowances that can be used in current or subsequent years. The United Nations Global Environmental Protection Agency would best facilitate trading by delegating the organization of spot markets to the private sector. Carbon dioxide spot markets could be established worldwide on existing organized spot markets, enhancing competition and efficiency in the marketplace. This would also include any designated initial auctions designed to encourage liquidity at the outset of the program.

Current contract markets and exchanges already have an infrastructure in place to facilitate electronic and open outcry trading. Finally, markets in participating nations need to demonstrate that they are sufficiently regulated. Transaction documentation should also be standardized.

A global warming treaty should include enabling language to ensure that less developed countries are subsidized to start spot and future markets in tradable entitlements. Other new markets have been subsidized in emerging markets. For example, the U.S. State Department has given a grant to the Chicago Board of Trade to assist the Budapest Commodity Exchange in developing its corn futures market.

Once the infrastructure for tradable carbon dioxide entitlements is in place in less developed countries, it could be extended to other commodities, financial instruments, and equities. Economists and international organizations have often focused on investments in capital equipment or major projects, such as dams, rather than establishing financial institutions, such as exchanges. The investment in exchanges and the establishment of programs similar to the state agricultural extension programs that teach the marketing and pricing of commodities could be of significant value. In fact, the applications of such marketing technology may be as important as efforts to improve agricultural productivity through irrigation projects and fertilization and harvesting technologies.

GLOBAL WARMING FUTURES

Although the sulfur dioxide allowance market is flourishing, there is as yet no historical evidence of a successful futures market in pollution property rights or of futures on tradable entitlements for emissions. There is, however, every reason to believe that these rights would lend themselves to trading in futures. Pollution allowance futures will require many of the same elements as other futures.

The role of a futures market is to provide a mechanism for price discovery as well as for hedging. Futures markets also may provide an alternative source of supply of the commodity. In order to capture these benefits, economists and market practitioners have defined a set of necessary conditions for a successful futures market. These include: (1) homogeneity, (2) existence of a spot market, (3) competitively determined prices, (4) price variability, (5) inefficient hedging alternatives, and (6) commercially acceptable contract design.

A cursory examination of the proposed market in CO_2 entitlements suggests that these conditions are easily satisfied. Homogeneity and competitive markets should be ensured if the signatories to the treaty include nations emitting a total of at least 20% of our planet's emissions. If signatories include the United States, the European Economic Community, and Japan, this would seem to be more than sufficient. Provisions should be made in each of these political units to transfer the entitlements to the private sector. This would ensure that no one country would hold an amount of entitlements sufficient to have a price-congestive effect on the market.

Alternatively, restrictions could prevent any individual entity from having more tradable entitlements than that of any one signatory, regardless of the base-year levels discussed earlier. This may provide a solution to both the perception and the reality concerning price congestion. Only time will determine whether price volatility will be sufficient to require hedging. Examination of the related markets in energy suggests that volatility will be present. Finally, once the spot market begins to mature, the viability of a futures market as a hedging mechanism should become clear.

DESIGNING A FUTURES CONTRACT

The final necessary condition for a successful futures contract is a contract design that meets the requirements of both buyers and sellers. The specific design of a CO_2 futures contract is patterned after the proposed SO_2 futures contract at the Chicago Board of Trade; see figure 3.2.1 (Sandor 1991).

---

1. *Unit of trading:* The contract unit shall be one-ton sulfur dioxide emission allowances.
2. *Standards:* Deliverable allowances shall be issued by the United States Environmental Protection Agency (EPA) and registered by the Board of Trade Clearing Corporation (BOTCC). Deliverable allowances shall be applicable against emissions in the current and future years.
3. *Months traded in:* Trading in sulfur dioxide allowance futures may be conducted in the current month and the next 11 months, plus quarterly listings in the March, June, September, and December cycle for up to three years from the current month.
4. *Price basis:* Minimum price fluctuations shall be in multiples of $1 per allowance.
5. *Hours of trading:* Hours for trading shall be 8:50 a.m. to 2:10 P.M. Central Standard Time.
6. *Trading limits:* Daily trading limits shall be $100 per allowance or $2,500 per contract. Variable limits shall be $150 per allowance or $3,750 per contract. No limits shall be in effect for the current month.
7. *Last day of trading:* No trades in sulfur dioxide allowance futures deliverable in the current month shall be made during the last three business days of the month.
8. *Delivery:* Delivery shall be by book entry on the ownership records of the BOTCC Corporation Sulfur Dioxide Allowance Depository.
9. *Position limits:* Speculative position limits shall be 2000 contracts in the current month and 5000 contracts overall.

Figure 3.2.1. Salient features of the Chicago Board of Trade clean air allowance futures.

Innovation in the futures industry typically begins with a hypothetical set of salient features of a specific contract. Subsequent parts of the process include modification of the initial salient features based on further research into the needs of hedgers and speculators (Sandor and Sosin 1983). It is in this spirit that the salient features of a futures contract in CO_2 are proposed in figure 3.2.2. The relationship between the two contracts will be explained later.

At this point, it is appropriate to analyze the initial salient features of the contract. Examination of both figures reveals that the CO_2 contract is based directly on the SO_2 contract. The architecture of the contracts is the same, but details will be different. Let's consider the trading unit or size of the contract. How and why the trading unit was established illustrates the thought process that leads to the creation of the contract. The SO_2 unit of 25 one-ton entitlements was based, to some extent, on forecasts that the value of allowances would range from $200 to $1,000 per ton. At an allow-

---

1. *Trading unit:* The contract unit shall be 10 one-ton tradable entitlements in carbon dioxide.
2. *Standards:* Deliverable allowances shall be issued by the UNGEPA and registered by relevant exchanges. Allowances shall be applicable against emissions in the current calendar year and for future years.
3. *Price basis:* Minimum price fluctuations shall be in multiples of $0.25 per ton.
4. *Delivery:* Delivery shall be by book entry on the ownership records of the UNGEPA and records of relevant exchanges.
5. *Trading months:* Trading in tradable entitlements for carbon dioxide shall be conducted in the current month and the next 11 months, plus quarterly listings in the March, June, September, and December cycles for two additional years.
6. *Trading hours:* Hours for trading shall be 8:30 A.M. to 2:30 P.M. Greenwich Mean Time or determined locally by designated exchanges.
7. *Trading limits:* Daily trading limits shall be $25 per allowance or $2,500 per contract. Variable limits shall be set at 150% of the daily trading limit. No trading limits shall be in effect for the current month.
8. *Last day of trading:* No trades in entitlements in carbon dioxide futures deliverable in the current month shall be made during the last three business days of the month.
9. *Position limits:* Speculative position limits shall be 2000 contracts in the current month and 5000 contracts overall.

Figure 3.2.2. Salient features of a tradable contract for carbon dioxide entitlements: a theoretical futures contract.

ance price of $500 per ton, the contract would be valued at $12,500. With this range of per-ton prices, the contract would range in value from $5,000 to $25,000. This value is consistent with the size of energy contracts traded on organized futures markets and, incidentally, grain and oilseed futures contracts. New contracts should have a risk-adjusted size so that speculative margins are consistent with other contracts. In this way, the new products can compete for speculative liquidity. Market experts suggested that the marginal cost of abatement for CO$_2$ might be 25% of that for SO$_2$. Therefore, following a similar line of reasoning, the trading unit was multiplied by four, and 100 one-ton entitlements were chosen as the initial unit.

The contract standards are a direct function of the way we define the commodity. It is important to emphasize that only those entitlements that are issued by the UNGEPA can be delivered on the respective exchange.

The minimum price fluctuation of $0.25 per ton or $25.00 per contract was chosen as a balance between maximizing the number of entry points for hedgers and speculators while also providing a large enough increment to attract market makers on the floor of the exchange (often termed "locals"). The $25 per contract lies between the $12.50 for grains and oilseeds and the $32.25 that is the typical minimum price increment for financial futures traded at the Chicago Board of Trade.

The delivery term is analogous to the mechanism provided by the CAAA for SO$_2$ allowances and is consistent with the proposed formation of the UNGEPA.

Trading month are identical to those of SO$_2$. The latter were chosen so that they could be arbitraged against the energy contracts that are traded around the world. There is obviously a similar link between CO$_2$ emissions and energy.

The choice of hours follows the SO$_2$ allowance futures contract. Both the SO$_2$ and CO$_2$ trading hours were chosen to maximize liquidity. The grain hours are from 9:30 to 1:15, and financial futures trade from 7:20 to 2:15. Therefore, floor traders (locals) from both of these markets could participate in the opening and closing periods of the environmental markets when most trading is likely to occur.

Trading limits, the last day of trading, and position limits follow the financial and agricultural markets. Using similar procedures minimizes the operational and regulatory costs of transacting on a new market.

Although one of the principal objectives of this paper was to draft a futures contract for CO$_2$ elements, it is a simple extension to write the rules and regulations of an organized options market. That contract would follow the same traditions used in drafting options on financial futures and grains and oilseeds.

Momentum has been building for specific actions to reduce global warming after the United Nations Earth Summit held in 1992 in Rio de Janeiro. Although joint implementation in specific instances, such as environmental programs in Costa Rica and the Czech Republic, is very promising, the total worldwide effort has been minimal. There has been little noticeable progress on other fronts, with the exception of the January 1995 United Nations Conference in Trade and Development (UNCTAD) Workshop in Prague. At that time, the author presented a simple, 10-stage program to implement the commitments made in Rio (Sandor 1994). These stages were:

1. Legally define an international property right for CO_2 emissions.
2. Designate and structure a global environmental protection agency.
3. Specify allocations and ensure that monitoring is in place.
4. Issue allocations and provide a homogeneous issuer.
5. Eliminate segmentation of value based on credit differences; for example, Costa Rica's allowances should not trade at a discount or premium to Germany's allowances.
6. Establish a clearinghouse — an easy task: Link international commodity exchanges and develop auction procedures.
7. Standardize contracts just as is done in the futures and swaps markets.
8. Provide for information sharing and dissemination of data.
9. Develop uniform tax and accounting conventions.
10. Subsidize exchanges in emerging markets to implement CO_2 trading. This will have spillover benefits for the marketing of commodities such as coffee and cocoa, thereby addressing the heart of many developmental issues.

The final report of the Prague Workshop advocates the idea of launching a pilot scheme to trade CO_2 emission permits to implement this 10-stage program (see Appendix A). A voluntary, nonbinding pilot program now appears to be gathering significant support. A subsequent meeting of experts on "Financial Issues of Agenda 21" took place in Glen Cove, New York, February 15–17, 1995. The participants also voted to endorse a voluntary, nonbinding pilot scheme in tradable CO_2 entitlements (see Appendix B).

The potential size of a tradable CO_2 emission credit market appears substantial enough to offer participants liquidity and flexibility to achieve stated stabilization goals. The tradable CO_2 emission credit market has the potential to be comparable in size to current

U.S. supplies of commodities such as wheat and soybeans that underlie the successful futures contracts traded at the Chicago Board of Trade. A start date of January 2000 seems very feasible, and the ability to achieve 1990 levels of emissions by the year 2010 is reasonable.

Unrelenting population growth and continuing industrialization of the planet ensure that the allocation of scarce resources such as air and water will be the most significant problem facing the twenty-first century and beyond. The proposals in this paper have been made in an effort to address and alleviate those problems. Further research into market-based solutions to critical environmental problems must be undertaken. Air and water are simply the planet's most important commodities.

REFERENCES

Blau, Gerda. 1944–1945. "Some Aspects of the Theory of Futures Trading." *The Review of Economic Studies* 8(1):1–30.

Coase, Ronald H. 1992. "The Institutional Structure of Production." *The American Economic Review* 82(4):713–719. September.

Grubb, Michael, and James K. Sebenus. 1992. "Participation, Allocation, and Adaptability in International Tradeable Emission Permit Systems." *Climate Change: Designing a Tradeable Permit System*, Chap. 9, "OECD Documents." Paris: Organization for Economic Cooperation and Development.

Sandor, Richard L. 1973. "Innovation by an Exchange: A Case Study of the Development of the Plywood Futures Contract." *Journal of Law and Economics* 16(1):119–136. April.

Sandor, Richard L. 1991. "CBOT Proposes Clean Air Futures for Emission Allowance Risk Management." *Financial Exchange* 10(2):9–10. September–October.

Sandor, Richard L. 1992a. "Implementation Issues: Market Architecture and the Tradeable Instrument (In Search of the Trees)." In *Combating Global Warming: Study on a Global System of Tradeable Carbon Emission Entitlements*." Geneva: United Nations Conference on Trade and Development, pp. 151–164, Geneva.

Sandor, Richard L. 1992b. "Environmental Futures." *Institutional Investor Forum*, Supplement to *Institutional Investor*, pp. 23–24. December.

Sandor, Richard L. 1994. "Model Rules and Regulations for a Global CO$_2$ Emissions Credit Market." Geneva: United Nations Conference on Trade and Development, pp. 61–105.

Sandor, Richard L., and Howard B. Sosin. 1983. "Inventive Activity in Futures Market: A Case Study of the Development of the First Interest Rate Futures Market." *Futures Markets*, edited by Manfred Streit. European University Institute, pp. 255–257.

Tietenberg, Thomas H. 1993. "Implementation Issues." *Study on Tradeable Emission Carbon Emission Entitlements*, Chap. 7. United Nations Conference on Trade and Development, Geneva.

A

EXCERPTS FROM PROCEEDINGS (CHAIRMAN'S) REPORT, WORKSHOP ON ECONOMIC INSTRUMENTS FOR SUSTAINABLE DEVELOPMENT, PRUHONICE, CZECH REPUBLIC, 12–14 JANUARY 1995

1 INTRODUCTION

1.1 Background

❏ The Workshop on Economic Instruments for Sustainable Development, held at Pruhonice, near Prague in the Czech Republic, between 12 and 14 January 1995, was set up at the instigation of the Czech government. The workshop was held under the auspices of the UN Commission for Sustainable Development (UNCSD) and a report was made to the UNCSD Financial Working Group (FWG) meeting held during March 1995 in New York.

The main purpose of the workshop is to initiate and stimulate international discussion on the use of economic instruments (EIs) as mechanisms and tools for environmental management. The workshop is expected to produce findings and recommendations to be passed on to the FWG for their consideration. The FWG will then prepare proposals for international demonstration projects, which can be implemented to examine and refine the practical use of economic instruments. These proposals will then be presented to the UNCSD.

This paper, representing the proceedings of the workshop, contains the main findings and recommendations (Section 2) of the workshop. These were presented to the delegates for their approval and adoption as the official report for submission to the UNCSD during the final session of the workshop.

1.2 Workshop Participation

Eighty-eight representatives from 23 countries, various UN organizations, commercial enterprises, financial and academic institutions, and nongovernmental organizations (NGOs) attended the workshop. Of the 21 countries represented, 11 countries were from OECD member states, 8 from Eastern and Central Europe, 2 from the Far East and the Pacific Rim, 1 from South America, and 1 from Africa.

2.1.8 Existing trading experience could be applied to a nonbinding pilot program in international CO$_2$ emission trading within the context of relevant international conventions. The purpose of the pilot program would be to carry out research, collect information and gain experience in this area.

2.3.5 Promote new applications for existing instruments and the development of new and innovative instruments for sustainable development.

2.3.6 Encourage the implementation of a nonbinding and voluntary pilot program of international CO$_2$ emission trading.

2.4.5 Governments are invited to consider participation in, and, where appropriate, promote the implementation of, a nonbinding and voluntary pilot project for international CO$_2$ emission trading.

B

❑

EXCERPTS
Report of the Second Expert Group
Meeting on Financial Issues of Agenda 21
Harrison Conference Center
Glen Cove, New York
15–17 February 1995

OVERVIEW

❑ 1. The second expert meeting on financial issues of Rio Earth Summit's Agenda 21, sponsored by the governments of Japan and Malaysia in collaboration with the United Nations Department for Policy Coordination and Sustainable Development (UN/DPCSD) and the United Nations Development Program (UNDP), took place in the Harrison Conference Center, Glen Cove, New York on 15–17 February 1995. The goal of the meeting was to continue discussions of financial issues of Agenda 21 and sustainable development started during the first such meeting (Kuala Lumpur, Malaysia, 2–4 February 1994) and to provide an input to the upcoming deliberations of the ad hoc intersessional working group on finance of the United Nations Commission on Sustainable Development (6–9 March 1995).

2. The meeting, attended by more than 40 experts from governments, international organizations, research institutions, and the private sector, was chaired by Dr. Lin See Yan of Malaysia.

3. The experts' discussion focused on the urgent need to close the

resource gap, that is, preventing full implementation of Agenda 21. Unless prompt and effective action is taken, this gap will continue to widen as the world economy diverges further and further from a sustainable path.

4. This report does not attempt to reflect all the views and suggestions made and does not represent a negotiated text. It reflects, nevertheless, the general thrust of the discussion. Participants agree that it provides a substantive contribution to future discussions by the Commission on Sustainable Development (CSD) and its ad hoc International Working Group on Finance.

B. Internationally Tradable CO$_2$ Permits

24. Internationally tradable CO$_2$ permits are a promising mechanism for bringing about cost-effective reductions in emissions of this greenhouse gas and transferring new and additional financial resources to developing countries and countries in transition. Experience with existing SO$_2$ markets in the United States is encouraging and should be examined for lessons that could be applied to CO$_2$ in an international context.

25. The CSD should encourage further study, with a focus on the development of a pilot scheme at an appropriate time, to gain experience and provide a framework for further research and development. Such a scheme should involve both industrial and developing countries on a voluntary basis in order to create opportunities for gains from trade and to ensure that developing countries can benefit from these opportunities from the outset.

26. Governments could encourage and facilitate the participation of emitters of CO$_2$ in particular utilities and heavy industries. The pilot schemes should be designed to assist governments participating in the Framework Convention on Climate Change to develop and implement effective mechanisms for dealing with climate change.

27. Experience with other traded goods suggests that commodity exchanges and other forms of market infrastructure would play an important role in developing markets for CO$_2$ permits. The CSD should encourage the active involvement of such parties in a pilot scheme for CO$_2$ permits.

Appendix

C

❑

ACKNOWLEDGMENTS

❑ I would like to thank Frank Joshua of UNCTAD for his intellectual support and incredible leadership. Penya Sandor contributed invaluable assistance in researching and editing this paper. Mary Kelly provided assistance in developing the original paper. Joseph Cole also provided valuable input. Margery Mandell's advice and editorial input were excellent. And special thanks to Marilyn Grace. I would also like to thank Richard F. Kosobud and the Editorial Board of the Workshop on Market-Based Approaches to Environmental Policy and Michael J. Walsh for their suggestions and recommendations. I would like to dedicate this study to Ronald C. Coase of the University of Chicago. His seminal article on social costs has clearly inspired tradable entitlements. On a more personal note, his advice and guidance on an article of mine that was published in the *Journal of Law and Economics* (Sandor 1973) inspired my later work in financial and environmental markets.

3.3

❑

JOINT IMPLEMENTATION: FROM POLICY TO PRACTICE

Edward A. Helme and Janet-Ann Gille

❑ Joint implementation (JI) is a provision, included in the Framework Convention on Climate Change, that allows for two or more nations to jointly plan and implement a greenhouse gas reducing or offsetting project. Joint implementation is important environmentally for two principal reasons: (1) it provides an opportunity to select projects on a global basis that maximize both greenhouse gas reduction benefits and other environmental benefits such as air pollution reduction while minimizing cost, and (2) it creates incentives for developing countries as well as multinational companies to begin to evaluate potential investments through a climate-friendly lens. While the debate on how to establish the criteria and institutional capacity necessary to encourage joint implementation projects continues in the international community, the U.S. government has created new incentives for U.S. companies to develop joint implementation pilot projects now. The United States has taken an active role in joint implementation, establishing two complementary domestic programs that allow U.S. companies to measure, track, and score their net greenhouse gas reduction achievements. With a financial commitment by three U.S. utilities, the Center for Clean Air Policy has developed a fuel switching, cogeneration, and energy efficiency project in the city of Decin in the Czech Republic that is the first project to complete construction and begin operation. The Decin project provides an ideal test case for assessing the adequacy and potential impact of the criteria for the U.S. Initiative on Joint

Implementation (USIJI) as well as for the criteria prepared by the United Nations Secretariat for the Framework Convention on Climate Change (FCCC).

THE FRAMEWORK CONVENTION ON CLIMATE CHANGE

The FCCC was signed in Rio de Janeiro on June 13, 1992. Although the treaty does not set specific timetables and targets, it does establish a goal for developed countries to stabilize their emissions of greenhouse gases by the year 2000 at 1990 levels.

> ...parties shall adopt national policies and take corresponding measures on the mitigation of climate change, by limiting... anthropogenic emissions of greenhouse gases and protecting and enhancing... greenhouse gas sinks and reservoirs. These policies and measures will demonstrate that developed countries are taking the lead in modifying longer term trends in anthropogenic emissions... recognizing that the return by the end of the present decade to earlier levels of emissions... would contribute to such a modification... *FCC Article 4.2(a)*

Nations with economies in transition, which include Central and Eastern Europe and the former Soviet states, are given the flexibility to choose a baseline year between 1987 and 1990 when they ratify the treaty. Developing countries signing on to the treaty do not have specific emissions reduction or offset goals but are committed to taking into account climate change considerations in their "social, economic and environmental policies and actions..." *(FCCC Article 2.1(f))*

Another key provision of the FCCC, and the focus of this paper, is joint implementation. Joint implementation allows two or more nations jointly to plan and carry out a greenhouse gas reduction project or a carbon sequestration project and to share the emissions reduction credit achieved by the project.

> ...Parties may implement... policies and measures jointly with other Parties and may assist other Parties in contributing to the achievement of the objective of the convention... *FCCC Article 4(2)(a).*

THE PILOT PHASE OF JOINT IMPLEMENTATION

The first Conference of Parties (COP) to the FCCC in April 1995 established a pilot phase for Joint Implementation (also known as activities implemented jointly, or AIJ). The pilot phase permits all nations to participate although it does not permit the transfer of credits for mission reductions achieved during this phase. The COP will decide whether to award credits for pilot phase projects by the year 1998 or 2000 at the latest, when the phase is scheduled to be completed. At present, participating nations are tracking, scoring, and reporting on JI projects to the FCCC Secretariat. At the second Conference of the Parties held in July 1996 in Geneva, the parties agreed to develop a consistent format for reporting and recording projects.

ONGOING JI DEBATE BETWEEN DEVELOPED AND DEVELOPING COUNTRIES

Annex I countries have agreed to establish greenhouse gas emissions baseline years (1990 for developed countries, chosen years for nations with economies in transition) and develop strategies to stabilize emissions, while non–Annex I countries (signatories to the treaty not included on the Annex I list—primarily developing countries) have no requirements to do either.

There is continuing concern from developing countries over the use of bi- and multilateral funding in joint implementation projects. The FCCC clearly commits industrialized countries to financially assist developing countries with "incremental costs" of environmental projects. Some potential host country governments are concerned that most or all assistance to non–Annex I countries will be considered joint implementation, bringing with it other obligations, most important, the transfer of the reductions achieved. By transferring the greenhouse gas reductions achieved to the investor country, some countries believe that they would be doing all the work for industrialized countries while making it more difficult for themselves if and when they chose to adopt baselines and target emissions levels. How greenhouse gas emissions reductions are to be shared between host and investor countries remains a JI feature critical to its success but yet to be resolved in general.

There was also concern that the potential for JI would weaken the commitment of industrialized countries to the Global Environment Facility (GEF), the likely conduit for multilateral aid for greenhouse gas–reducing projects under the FCCC. Supporting nations agreed in 1995 to replenish the GEF at approximately $2 billion ($220 million for Central and Eastern Europe and the former Soviet states), but U.S.

budget cuts have eroded U.S. support and weakened support among other Annex I countries.

While the G-77 coalition of developing countries and China continue to argue for more GEF-type assistance from developed countries, individual countries are seeking ways to develop JI to benefit their local environment. A growing number of non–Annex I countries are keenly interested in exploring the potential for international investment in their countries, both to benefit the environment and to expand economic development. One such country is Costa Rica, the current chair of the G-77.

Costa Rican public and private leaders have recognized the potential benefits of JI to furthering these goals. Like all nations, the costs of meeting environmental goals in Costa Rica are a major hurdle to overcome.

Already by 1995 Costa Rica had:

- Developed its own JI program, which has identified and characterized a variety of greenhouse gas reduction and carbon sequestration projects — many of which have received U.S. Initiative for Joint Implementation approval.
- Founded the Costa Rican Office for Sustainable Development to coordinate all sustainable development projects within the country and promote these principles internationally.
- Established the first-ever bilateral agreement with the United States on joint implementation.
- Developed the concept of certified tradable offsets (CTOs), which assure investors of the quality of the carbon offsets they are purchasing.
- Held several international conferences, both to learn more about joint implementation and to update interested parties on their progress in developing and implementing sustainable development goals.

OPPORTUNITIES IN ANNEX I COUNTRIES

Significant opportunities are available for joint implementation between Annex I countries. Internationally, financial resources are scarce — from individual industrialized countries, as well as multilateral funding sources — for assisting nations whose economies are in transition in dealing with their economic problems, let alone their environmental ones. But, the need for environmental enhancement in these transitional nations is immense. Years of energy inefficiency and disregard for environmental issues have left these countries with not only large greenhouse gas emissions but also dangerously high levels of SO_2, NO_x, particulates, and air toxics. It has also left them with relatively inexpensive ways to reduce these emissions if capital and technology can be made available. For these nations, joint implementation offers both enhanced air quality and improved health for their citizens, and this has led these countries to eagerly support the JI concept on the COP.

Just a sample of the projects that are already being developed gives a sense of the range of opportunities for environmentally and economically attractive JI projects.

- The potential for wind energy in Ukraine has already spurred a joint project between Kenetech, a U.S. wind products and service company, and Krimenergo, a Ukrainian electric utility. Together they will not only provide 500 MW of wind energy to the region but will also assist Ukraine to convert from defense industries to wind energy equipment production.
- Norway, working with several partners, has begun a project in Poland to assist in the conversion of coal-fired boilers to natural gas. The conversions are likely to include small cogeneration schemes as well as gas-fired high-efficiency condensing boilers.
- Other project opportunities include working to reduce leakages on gas pipelines throughout Russia; coal-bed methane recovery projects throughout the region; and fuel-switching and energy efficiency projects in cities and towns in Central and Eastern Europe. Leading the way, we believe, is the Decin model.

THE DECIN MODEL

Background

Located in the northwestern corner of Northern Bohemia in the Czech Republic, Decin is a heavily industrialized center with a population of 55,000. The city sits astride the Elbe River in a deep valley about 10 km wide, at the base of ridges that are 500 meters tall. Eighty percent of Decin's residents live in multifamily flats.

Most of the city's housing units are heated by local brown coal—used to fuel the district heating systems and individual homes. As of 1992, 75% of the square footage in the community was heated by coal, with about half of this total linked to the district heating system. Because the stacks on the district heating plants are low, emissions from the plants stay within the valley. The extensive use of brown coal is the principal reason for the high levels of air pollution in the region—from sulfur dioxide (SO_2), nitrogen oxide (NO_x), air toxics, and particulates. While large power plants within the Czech Republic have begun to receive assistance, district heating plant projects are often too small to attract multilateral funding despite the devastating effect these plants have on the area.

Because of the mountains and deep valleys and its heavy reliance on brown coal, Northern Bohemia is one of the most polluted regions in the Czech Republic. Concentrations of some air pollutants far exceed national and international health standards. For example, the highest 24-hour average for SO_2 concentrations occurred in January 1987 and was more than 10 times higher than health standards set by the World Health Organization. Heavily contaminated air threatens human health throughout the region. Childhood mortality rates are twice that of the rest of the Czech Republic. Half of the pediatric patients in area hospitals suffer from conditions linked to air pollution. And, on average, life expectancy for residents in the region is five years shorter than in other parts of the Czech Republic. To prevent the large-scale relocation of environmental refugees, the former regime resorted to financial compensation—termed "burial money" by recipients—to those who stayed in the region.

Under the leadership of Mayor Kunc, Decin has placed the highest priority on developing specific projects to begin immediately reducing air pollution in order to improve the health of local residents. In the last several years, the city has become a leader in the Czech Republic for its efforts in addressing its environmental problems. A year-long environmental study, completed for the city by the Danish consulting firm of Bruun and Sørensen in April 1993, provided recommendations for the city to address its local air quality problems. Options and recommendations offered in the study include:

- Revamping the district heating system—supply-side efficiency and fuel switching.
- Improving energy efficiency in apartment blocks.
- Installing energy control equipment in units.
- Switching heating systems in individual houses from coal to gas.
- Switching large industrial users from coal and/or heavy oil to efficient gas systems.

The Project

With financial assistance from three U.S. utilities — Wisconsin Electric Power Company, NIPSCo Industries of Indiana, and Commonwealth Edison of Illinois — the Center for Clean Air Policy has developed with the city of Decin a first-of-its-kind pilot joint implementation project. One of the city's five district heating plants, the Bynov plant, which provides heat and hot water to multifamily dwellings and several businesses, has been converted to a gas-fired cogeneration facility. At the same time, energy efficiency improvements have been made to the heat distribution network, allowing the facility to be downsized from 19.6 to 10.6 MW.

Previously, the Bynov plant had three 5.6-MW coal-fired boilers and one 2.8-MW coal-fired boiler. The existing natural gas pipeline has been extended to connect to the district heating plant. With partial financing from the three U.S. utilities, the city has replaced existing coal-fired boilers with new, more efficient gas-fired engines and a gas-fired peak boiler. The project has also replaced the existing, poorly insulated pipes that carry steam to the multifamily dwellings. In the past, the plant lost great amounts of efficiency by converting the hot water to steam at the plant, then recondensing the steam back to hot water at the dwellings. With the new, well-insulated pipes, the project eliminates the need for conversion from hot water to steam.

Impacts of the Project

The fuel switch, cogeneration, and efficiency improvements will reduce air pollution, with benefits to both the regional and global environment. The measures being invested in by the three U.S. utilities and the city will have the following impact:

- Carbon dioxide: reduced by 5991 tons/yr (carbon weight) on-site from use of gas for heating purposes and an additional estimated 20,000 tons/yr off-site from the reduction in use of existing coal-fired plants. The on-site reduction is about one-third from baseline emissions.
- Sulfur dioxide: reduced by 96 tons/yr (virtually eliminating SO_2 emissions from the district heating plant).
- Ash: reduced by 3190 tons/yr (total elimination of the waste stream).
- Particulates: completely eliminated.

While the CO_2 reductions will help to reduce the threat of global climate change, the immediate results from the project will be felt most directly in the reduced levels of SO_2, NO_x, air toxics, and particulates. For the community of Decin, this project will continue to improve the air quality and to alleviate the health problems that have resulted from uncontrolled emissions of air pollutants.

Financing the Project

In attempting to finance this project, the city of Decin found that, if they received a loan, they would have to pay interest rates in the high teens and pay back the loan, on average, in four to eight years. If they could increase the project downpayment, the interest rates would decrease and the payback period would increase. Without foreign investment, however, they would probably be unable to increase the downpayment and, thus, be unable to finance the project at the prevailing rates.

The city was considering two options for the Bynov plant—a simple fuel switch with a capital cost of $1.5 million and a cogeneration option with a capital cost of $8 million. Because of the high capital cost of cogeneration, the city initially focused on the simple fuel switch while still considering cogeneration. The city initiated discussions with the Severoceska energetika a.s .(SCE)—one of eight national electricity distribution companies—and received an agreement to buy power from the potential cogeneration project. The larger project, especially with the commitment from SCE to purchase the cogenerated power, was much more attractive to investors.

That's where joint implementation comes in. The three U.S. utilities will provide a 25-year $600,000 no-interest loan to the city which, in turn, will use the capital to finance the remaining portion of the project costs. Once this funding was secured, the city decided to seek funding for the full cogeneration project rather than the simple fuel switch. Because of the U.S. utilities' financial commitment, the city was able to leverage financing from the Czech State Environmental Fund and obtain a grant from the Danish Ministry of Environment for the balance of the project costs. Without this commitment, it is unlikely that the city would have been able to fund either the simple fuel switch or the cogeneration project.

The cost to the investors of the CO_2 reductions is estimated to be approximately $4 per ton of carbon dioxide. This calculation is based on the fact that the U.S. utilities are providing a 25-year no-interest loan to the city in return for 5991 tons of the CO_2 reductions achieved on-site annually by the project. The additional off-site reductions, approximately 20,000 tons, will remain in the baseline of the Czech Republic, contributing to that country's commitment to stabilize emissions under the FCCC.

The potential for replicating this first-of-its-kind private sector JI project in other Central and Eastern European communities is enormous. There are an estimated 5000 coal-fired district heating system or industrial boilers in the Czech Republic, 15,000 in Poland, 40,000 in Russia, 16,000 in Ukraine, and 3500 in Hungary. Many are outdated and inefficient and operate with minimal pollution controls.

U.S. DOMESTIC PROGRAMS AND JOINT IMPLEMENTATION

What does all this mean for U.S. companies, like those investing in the Decin project, that are seeking to invest in JI projects? Completion of the international process required to establish the criteria, institutional capacity, and authorization of credits necessary to produce incentives for JI will be several years down the road. The international process is time-consuming, but this does not need to diminish the interest of U.S. companies in JI. The United States has already put in place two domestic programs that will ensure that their efforts are "measured, tracked and scored." In November of 1992, the Energy Policy Act of 1992 was enacted, establishing the first registry of U.S. greenhouse gas emissions through the National Inventory and Voluntary Reporting of Greenhouse Gas Emissions (Section 1605(b)). Less than a year later, the U.S. Climate Change Action Plan (U.S. CCAP) was announced; this includes the U.S. Initiative on Joint Implementation (USIJI), a U.S. pilot program for JI.

Both these programs offer U.S. companies the incentive to move ahead with JI projects and the expectations that they will be "rewarded" for their early action. In addition, the U.S. government's support for a "realistic, binding medium-term target" for emissions reductions, enunciated at the second Conference of Parties in July 1996, has created added interest in JI. The United States made clear, in its statement at that conference, that emissions trading and JI should be key elements of this new international target.

THE NATIONAL INVENTORY AND VOLUNTARY REPORTING OF GREENHOUSE GAS EMISSIONS

Section 1605(b) of the Energy Policy Act of 1992 authorizes the Department of Energy to develop a registry to record impacts of voluntary efforts by the private sector to address net greenhouse gas emissions. The registry is to include measures undertaken both in the United States and in other countries. For example, the U.S. CCAP refers to Section 1605(b) as the mechanism that "companies participating in the USIJI will use to provide information on the progress of overseas projects..."

The program is designed to provide U.S. industry with the opportunity to voluntarily record its efforts to reduce emissions at the source and to offset emissions through carbon sequestration (forestry and agricultural) projects or through off-site activities, such as landfill methane recovery, fuel switching, or energy efficiency projects either in the U.S. or internationally. The program creates a database to record these initiatives and their verified net emissions benefits. This information is expected to be used at a later date, first, to track U.S. efforts to meet our goal of stabilizing net greenhouse gas emissions at 1990 levels by 2000 and, second, to credit and recognize those forward-looking industries that take the initiative now to reduce greenhouse gas (GHG) emissions and voluntarily record those net reductions under the program. Some U.S. industries have been interested in acting now to reduce emissions but have been concerned that these early efforts would not be recognized. This voluntary recording provision was designed to allay those fears and encourage U.S. industry to be proactive in reducing emissions. To accomplish that, it is extremely important that the registry be designed to evaluate projects based on their credibility and accuracy. Final guidelines for implementing the registry were issued in October 1994.

LESSONS FROM THE ACID RAIN DEBATE

There is concern on the part of some potential U.S. participants in JI (or even in other domestic activities to reduce GHG emissions) that, in effect, their "good deed will not go unpunished." Some U.S. utilities remember difficulties that arose, during the Clean Air Act Amendment debates, for utilities that wished to receive credit for actions they had taken to reduce sulfur dioxide emissions early.

Given the uncertainty involved in crediting JI projects, why would U.S. companies move ahead? There are two reasons. First, as a new concept, JI pilot projects must demonstrate the environmental and economic benefits that can be achieved through implementation. Second, the voluntary registry established under Section 1605(b) can provide sufficient incentive for companies to become involved in these projects. The lessons learned during the acid rain debate can be instructive.

Under early versions of the acid rain bill, utilities that had taken early action were getting a much smaller number of allowances than the utilities that had done nothing. It seemed that the fewer measures taken, the greater the reward. In the end, for utilities that had significantly reduced emissions prior to the baseline years, a provision was added that provided them 120% of their allowance allocations. At the same time, allowance allocations to the utilities that had not taken early measures were ratcheted down. This ensured that the 8.9 million ton cap was unaffected while allowances were allocated on a more equitable basis.

This exact same process can work for companies that register projects under Section 1605(b) of the Energy Policy Act (EPACT)— domestic and international. To ensure that the companies that acted early to reduce or offset emissions of greenhouse gases receive credit, the United States could develop a similar GHG allowance allocation system. For example, if the United States adopted a cap on greenhouse gas emissions reducing national emissions to 10% below 1990 levels, those well-measured and verified projects that are included on the database will be "credited" to the project participants. At the same time, the GHG allowances allocated to companies that did not take early action would have to be reduced by an equal amount. If, at that time, the international community still did not accept JI as a component of national greenhouse gas reduction plans, ongoing JI projects that are credited to U.S. participants would also have to be compensated for by increasing the emissions reduction burden on nonparticipating companies. It is likely that any international treaty to restrict CO_2 emissions further will leave allocation or reduction burdens within individual countries to individual national governments.

For example, take a very simple hypothetical national emissions inventory. The inventory includes three companies—A, B, and C— each emitting 100 ton/yr in 1990. Over the next 10 years, 1990–1999, companies A and B have growing baseline emissions; company C's emission level remains constant at 100 ton/yr. In 1995, company A establishes a joint implementation project that transfers 15 ton/yr of

GHG reductions from the host country to company A. The project and the emissions reductions are registered under USIJI and Section 1605(b).

After 1999, the country establishes an emissions cap of 240 ton/yr (a 20% reduction from 1990 levels)—80 tons for each company. At this point, actual emissions for the three companies are as follows: company A, 110 ton/yr; company B, 110 ton/yr; and company C, 100 ton/yr. Because of its JI project, company A also has 75 tons of offset reductions registered under Section 1605(b). If company A received credit for them, without any adjustments to the other companies, the emissions cap would be raised to 315 tons, above even a stabilization target.

Instead, to ensure that the cap is maintained, a strategy parallel to the legislative solution on acid rain would be to divide the 75 tons of offsets by the five years in which they were achieved and add that total to the first five years of company A's baseline. Therefore, company A's GHG allowance allocation is 95 per year for the first five years (2000–2004) instead of 80. At the same time, an equal amount—15 tons—must be deducted from the emissions baselines of companies B and C, reducing their GHG allowance allocations by 7.5 ton/yr. Their GHG allowance allocation would then be 72.5 per year for the first five years; and total allowance allocations for each year would remain at 240. In addition, company A will be allowed to continue to offset its emissions by 15 ton/yr through the ongoing JI project. If JI counts internationally, the reductions from companies B and C would be unnecessary. Either way, company A is a winner.

An international decision on a new reduction target is slated for December 1997 at the third Conference of Parties in Kyoto, Japan. While implementation of a mandatory cap and allowance system is still years away, the above example helps to explain how early reductions might be credited eventually at the national level. The key, however, to creating extensive interest in JI projects in the private sector will be an international decision in Kyoto on binding targets that relies on emissions trading and joint implementation.

U.S. INITIATIVE FOR JOINT IMPLEMENTATION

The development and implementation of the Decin project—and any other international projects that may seek registration under the U.S. CCAP—must be completed with full consideration given to USIJI eligibility criteria. The final rules and guidelines for the USIJI program were published in June 1994. Criteria for these projects include the following:

- Projects can be implemented in any country that has "signed, ratified or acceded to" the FCCC.
- The project must be formally accepted by the host country government.
- The project must yield net reductions in greenhouse gas emissions that are in addition to business as usual (i.e., the project was undertaken or revamped with joint implementation in mind).
- Federally funded projects must be undertaken with funds in excess of those available for such activities in fiscal year 1993.
- The data needed to estimate current and future net greenhouse gas emissions in the absence of, and as a result of, the project must be provided.
- There must be adequate assurance that actual net greenhouse gas reduction benefits accumulated over time will not be lost or reversed.
- The project must include adequate provisions for tracking actual emissions resulting from the project and external verification. Annual reports on actual net reductions in greenhouse gas emissions and on the share of such reductions attributed to each of the participants, domestic and foreign, must be provided to an interdepartmental evaluation panel.
- Other environmental impacts/benefits of the project must be identified.

The USIJI Evaluation Panel, comprising representatives from eight U.S. departments or agencies, has been established to: advise and assist project sponsors, hosts, and developers; accept, review, and evaluate projects; approve or reject projects; certify net emissions reductions estimates; and provide an annual report to include a summary of approved projects. In addition to the guidelines summarized above, the panel is to make determinations on whether to include projects under the USIJI after considering:

- The potential for the project to lead to net changes in greenhouse gas emissions elsewhere.
- The potential positive and negative effects of the project apart from its effects on net greenhouse gas emissions.

- Whether the U.S. participants are net emitters of greenhouse gases within the United States and, if so, whether they are taking measures to reduce such emissions.
- Whether efforts are under way within the host country to ratify or accede to the FCCC, to develop national inventories and/or baselines, and to adopt and implement mitigation measures.

The Decin project was one of the first seven to be approved by the U.S. IJI and is the first to complete construction and begin operation.

CONCLUSIONS

Since 1992, there has been significant movement in the global climate change debate. Because greenhouse gas emissions and emissions reductions know no national boundaries and because the impacts of global climate change threaten worldwide economies, this debate has been going on both in individual nations and in the international arena. A major focus of these discussions has been the potential for joint projects, between nations, to reduce greenhouse gas emissions. The potential for these projects, both in nations with economies in transition and in developing countries, is enormous, as is the need for environmental investment. In Central and Eastern Europe and the former Soviet states alone, the opportunities for investment in energy efficiency and cogeneration projects are significant.

The United States has promoted the concept of JI and provided incentives for companies to invest in JI projects. While the international debate continues, USIJI and Section 1605(b) of the Energy Policy Act offer U.S. companies the assurance that their efforts to reduce greenhouse gases today will not go unnoticed. By registering the results of these projects, we help move the international debate along by demonstrating the benefits that can be achieved by these projects — in reducing the threat of global climate change, enhancing the environment in host countries, and promoting the efficient use of economic resources. At the same time, in looking to the future, we realize that, when the international community chooses to set specific caps on greenhouse gas emissions, the registry can be used to ensure that these good deeds are rewarded, not punished.

REFERENCES

Anderson, Robert J. 1994. *Joint Implementation of Climate Change Measures: An Examination of Some Issues.* Washington, DC: World Bank, p. 5. January.
Clinton, Bill. 1993. Earth Day Address. April 21. Washington, DC.
U.S. Climate Change Action Plan. 1993. Washington, DC. pp. 4, 26. October.
U.S. Department of State. 1992. Publication 10026. Bureau of Oceans and International Environmental and Scientific Affairs, Office of Global Change, National Action Plan for Global Climate Change. December.

3.31

□

DISCUSSANT

Richard M. Peck, University of Illinois at Chicago

□ My intent is to provide another perspective on Dr. Sandor's talk. Houthakker's well-known paper "The Economic Function of Futures Markets" (1959) begins with the sentence "Economic analysis of institutions is not highly regarded or widely practiced among contemporary economists." Since Houthakker wrote that sentence in the 1950s, fashion has changed, and the economic analysis of institutions is more respectable. The new institutionalism stresses property rights, transaction costs, complete information, and agency costs. These concepts are powerful tools for understanding how economic systems change over time. Markets are primary institutions and are likely to rise and work well when property rights are well defined, transaction costs low, and the number of participants large. Transaction costs are low if the traded commodity is well defined and relatively homogeneous. If markets for CO_2 emission are to be viable, these conditions will most probably have to be satisfied.

Property rights are established when the incremental benefit of doing so exceeds the incremental costs. Property rights always require enforcement; if enforcement is very costly, then property rights will not be well established. Correspondingly, if the benefits to establishing property rights are very low, then property rights may not be established. This simple framework explains why serious proposals for marketable permits are occurring now and not 40 years ago. Tradable SO_2 permits have been available for several years, and permits related to urban smog have begun to be traded in California; serious proposals for CO_2 are on the table. Because of rising emission levels, the benefits of pollution markets are much higher than they used to be. In addition, technological changes, such as continuous emission monitoring, have reduced the cost of enforcing pollution property rights. With rising benefits and falling costs, the emergence of a system of property rights for emissions is inevitable.

How does a futures market fit into all this? The presence of an ongoing, vital futures market signals that the spot market is working well. Put differently, a futures market is a sufficient statistic for an efficient spot market. In a loose sense, the presence of a futures market is only a sufficient condition for an efficient spot market; that is, the absence of a futures market does not necessarily mean that the spot market is not operating efficiently. After all, there is no futures market for screwdrivers, but the spot market for screwdrivers is surely efficient. But, in the absence of a futures market, one may have a little less confidence that the spot market is functioning efficiently. Of course, the primary function of a futures market is to allocate risk from more risk-averse individuals to less risk-averse individuals, thereby improving the allocation of resources. In environmental markets, risks arising from sources such as governmental policy, technological change, and demand and supply shocks abound, suggesting an important allocative role for futures markets. Dr. Sandor's paper provides a nice perspective on what needs to be satisfied for a viable futures market to emerge and a look at the institutions that might be in store for us.

REFERENCES

Houthakker, H. 1959. "The Economic Function of Future Markets." In *The Allocation of Economic Resources*, edited by Moses Abramowitz. Stanford, CA: Stanford University Press.

3.32

❑

DISCUSSANT

Kevin G. Quinn

❑ Dr. Sandor's paper is a valuable contribution to the discussions surrounding the use of tradable emission permits for carbon dioxide control. His insights into the nuts and bolts of creating the financial-legal instruments meant to serve as these permits, and the markets in which they would trade, provide guidance on an extremely important facet of the application of the theory of market-based environmental policy. All too often, short shrift is given to "un-glamorous" mechanical issues until implementation is imminent, even though policy success or failure often turns on these issues.

An important consideration in the design of these instruments is to create conditions conducive to thick trading, with many buyers and sellers. The key to successful use of markets to achieve environmental goals is that emitters use market prices to guide their emission reduction choices. Speculators with little or no direct interest in CO_2 emissions can provide the liquidity necessary to encourage emitters to trust these price signals, allowing the invisible hand to perform its advertised magic. It is critical to realize that such speculators have a myriad of financial options; their participation will be based on the extent to which tradable permits are allowed to compete with these options. In the event that the U.S. government mandates national CO_2 emission reductions, Dr. Sandor's paper very effectively outlines how to encourage this speculative activity.

But should these markets be opened for trading by other countries' governments and economic agents? This is not an insignificant question. Billions of dollars globally might be saved if there were international emission trading. But are these savings truly possible? Can trading involving other countries work?

Suppose that global warming treaty that makes use of inter-country emission trading is indeed struck and emission rights (permits) are allocated to the countries of the world. What is to prevent some countries from cheating, that is, emitting without consuming a permit or, expressed differently, failing to back up a permit sold with a unit of emission reduction? Cheating is cancerous to emission permit markets; it depresses permit prices and sends incorrect signals to other emitters, thus diminishing the cost-effectiveness of the policy. Cheating also increases environmental damages and, if pervasive, may cause markets to collapse completely. The success of international trading depends on countries honoring the greenhouse treaty.

Countries generally adhere to the treaties they sign, despite the absence of a global police force with teeth swift and sure. This is because the cost of cheating—castigation by the community of nations—is sufficient to make governments honor their commitments. But economies in the late twentieth and early twenty-first century world live on fossil energy. Significant restrictions on the use of coal, petroleum, and natural gas are likely to have very high costs, at least in the short run as economies adjust. The pain of being labeled an environmental miscreant may pale beside the specter of deep emission cuts. Failure to take into account the temptation to cheat will almost certainly doom a global greenhouse policy based on CO_2 trading. The key to success is due consideration to the cost-benefit calculus facing governments.

Allocation of emission permits among countries may be the most important influence on the cost-benefit relationship. More permits must be allocated to countries facing the largest costs and the least benefits, that is, those with the greatest temptation to cheat. More permits mean greater permit sales revenues (or lower purchasing costs), in effect, a side payment that makes the policy more palatable. Cheating may be deterred even more by the prospect of having this side payment withdrawn, that is, of cheaters being barred from selling their excess permits.

Allocating permits with this in mind may be very different from allocations based on population, GDP, baseline emissions, or other equity-related criteria. Fairness alone is not sufficient to guarantee that one or more major emitters will not face too large a cost-benefit spread to participate. The success of an international CO_2 emission trading system may require that better-off countries or past "environmental villains" receive a large share of the total permits allocated.

Resolution of this issue may reduce the degree to which global emissions can be mitigated using trading and may even present an insurmountable hurdle.

If participation by other countries is deemed possible, how should this trading take place? In particular, should national governments be the sole conduit of international trading, or should foreign firms and citizens be allowed to engage in trades? Empowering individuals to trade will serve the cause of thick markets, but can an exchange clearinghouse manage to guarantee both ends of a trade when one of the parties is in a country where law enforcement is iffy? These are important questions; successful intercountry trading has the potential to save billions of dollars.

If the United States goes ahead with real reduction of CO_2 emissions from within its borders, traditional commodity exchange markets can and should play an important role, regardless of whether intercountry emission trading is a part of international climate change strategy. If managed properly by U.S. regulators and the exchanges, the resulting success might induce other countries to participate in trading. Movement toward full-scale international trading is likely to be evolutionary, however. We face a unique opportunity. Dr. Sandor's paper is a useful guide for U.S. policymakers thinking about using the market to effect greenhouse policy.

The paper by Helme and Gille examines the joint implementation (JI) provisions of the Framework Convention on Climate Change (FCCC). Pointing to the leadership that the U.S. government has offered under the Climate Change Action Plan (CCAP), the authors appear to be optimistic that JI provides an opportunity to harness market forces for service in reducing global greenhouse gas (GHG) emissions. I am not so sanguine. Although the authors furnish an informative identification and discussion of a number of important diplomatic questions that face JI, deeper economic issues remain. It is economics, not international diplomacy, that ultimately will decide JI's role in global climate change (GCC) policy.

The movement toward international cooperation on global climate change must start as a crawl before it can become a headlong run. Joint implementation might be this crawl, but its voluntary nature under CCAP means that its scope will be quite limited. At best, JI will be the launchpad for more effective market-based weapons in the global climate change arsenal. At worst, JI will become an example for other, long-standing and yet unsuccessful agendas, with little or no real effect on global GHG emission time paths. If the latter scenario more closely describes how JI programs actually are implemented, then the disillusionment that could result might hamper future GCC policy initiatives.

Joint implementation is predicated on the assumption that developing countries and countries just in the process of establishing markets have relatively inexpensive opportunities to reduce GHG emissions with respect to "no-policy" scenarios. Richer countries might fund projects that take advantage of these opportunities, reducing global emissions less expensively than by efforts within their own borders. This sets up a win-win situation: Poorer countries get funding for environmentally and developmentally friendly projects that would otherwise be impossible, and rich countries get credit for emission reductions at bargain rates. On the surface, this scenario echoes trading behavior in markets. Might the availability of JI, in effect, create an emission credit market? Elementary microeconomic analysis offers only a very tempered optimism.

Functioning markets require the following: a well-defined commodity that includes a set of property rights; sellers who profit from supplying the commodity; buyers who demand the commodity for consumption or as an input to a production process; and institutions through which buyers and sellers can transact trades without friction (or at least for relatively little cost). As yet, JI does not meet these requirements.

What exactly *is* a JI emission credit? The non–Annex I countries — those that presumably would have many of the credits to be sold — have no requirements under FCCC to come up with a benchmark by which to gauge future national emission levels. This is a problem: Credits are supposed to be emission reductions relative to these essentially indeterminate baseline (no-policy) levels. Even if the reductions can be quantified, how are they to be verified? Without quantification and verification, it is difficult to recognize JI emission credits as "commodities" that will be freely bought and sold.

The Climate Change Action Plan attempts to provide some guidance on this issue by enumerating rules for certification of JI credits, but certification under the CCAP is shaping up as extremely bureaucratic, byzantine, ad hoc, and *expensive*. Furthermore, there apparently is no real property right; a credit easily can be willed out of existence by government fiat. To claim that certification and registration under the CCAP constitutes "ownership" of a JI credit requires a somewhat pliable imagination.

How does a seller of a JI credit profit from the transaction? Despite the best efforts of negotiators to limit JI efforts to incremental aid increases, recipients of JI assistance have good reason to believe that they might be diminishing other aid sources, sources with fewer strings attached. Governments in developing and "marketizing" countries, always suspicious of the motives of richer countries, may not rush to participate in JI. Furthermore, the likely supply of JI emission reduction opportunities pales compared to the scale of the CO_2 emission problem. Consider the Decin project cited in the paper. At an annual rate of increase in baseline U.S. CO_2 emissions of 1% per year (a conservative estimate), it would take more than 10,000 Decins just to stabilize U.S. emissions for the 1995–2005 period.

The generators of JI credits are presumed to be firms and governments in developed countries. Why should a firm generate JI credits? These are essentially public relations efforts and hopes for future recognition of the CO_2 reduction. In the United States, the energy industry is already stressed by the inexorable charge toward competitive markets for power generation and distribution. I sincerely doubt that these firms are willing to commit the large sums required to reduce, even slightly, U.S. CO_2 emissions from baseline. The desire on the part of private firms in other developed countries to spend large sums on JI is not likely to be very different. Good public relations is simply not sufficient incentive to buy credits that may turn out to be worthless. Governments may not be able to fill this void as they currently face scarce foreign aid resources; in the United States, current levels are already under attack; increases for environmental or other "nonstrategic" purposes seem doubtful.

Finally, the transaction costs of JI projects seem prohibitive. Search and information costs, in particular, will be high relative to the emission savings; this would get worse as the easiest, best projects (the "low-hanging fruit") are exploited. The paper's analysis of the Decin project does not include an estimate of the resources required to wed the buyers and the sellers of the JI emission credits. My guess is that these costs were not insignificant compared to the $600,000 contributed by the three U.S. utilities. In general, international projects involving developing or countries developing markets are infamous for the bloated bureaucracies they engender. Markets do not work well in the face of high transaction costs.

Joint implementation, in fact, may achieve some reductions in global carbon emissions over the next 10 years or so and can even be a critical stepping-stone to more sophisticated and effective CO_2 emission reduction markets. Expectations regarding its potential should be realistic, however. The relief that JI can actually provide for greenhouse warming is likely to be modest.

3.4

❏

GREEN FEES: FISCAL INSTRUMENTS FOR SUSTAINABLE DEVELOPMENT

*Robert C. Repetto**

❏ Fears of lagging productivity growth and loss of international competitiveness in the U.S. economy have led to concern about the costs of meeting challenges to the quality of life and the environment. The policy choice has been cast as one between environmental protection and jobs or income. The conflict is not limited to Washington, DC. In the nation's cities and states, revenue deficits are impairing the ability of authorities to respond to deteriorating physical and social infrastructure and acute urban problems.

The resources with which to address these domestic and international problems are not at hand. Americans already feel burdened by taxes. The typical family is struggling to maintain living standards. Between 1971 and 1990, median family income adjusted for inflation rose only from $33,191 to $34,213, a gain of only 3%. For families below the median, the gains, if any, were even smaller. Economic improvements over these years were achieved almost entirely through increased work effort, mostly by women, as the civilian labor force participation rate rose from 60% to 66%. For those employed, average hourly inflation-adjusted earnings in private employment fell from $8.53 in 1972 to $7.46 in 1991, and average weekly earnings declined from $315 to $256. Compared to 10 or even 20 years ago, the American population is working harder and making less.

*Robert Repetto is vice president and director of the Economics Program at World Resources Institute. This paper is based on a WRI report (Repetto et al. 1992).

At the same time, Uncle Sam is taking more out of the average person's paycheck. Between 1972 and 1990, personal income taxes and Social Security payroll taxes together have risen from 17.5% to 19.2% of personal income. Largely because of the rise in payroll taxes, most people have not seen any cut in the taxes they pay. There is a widespread view that tax cuts are only for those with high-priced lawyers, accountants, and lobbyists.

For these reasons, taxes have become extremely controversial. Tax policy is highly sensitive, but political debate has dealt only with *how much* we tax, not *what* we tax. This is unfortunate. What we tax is important. At present, our taxes fall mostly on just those activities that make the economy productive: work, savings, investment, and risk taking. Naturally, such taxes discourage people from undertaking these vital activities. A better system would place more of the tax burden on activities that make the economy unproductive and that should be discouraged: resource waste, pollution, and congestion, for example. Taxes on these environmentally damaging activities would not distort economic decisions but, rather, would correct existing distortions.

Almost all taxes have incentive and disincentive effects. Although economists talk of taxes that don't affect behavior, "lump-sum taxes," there are almost no practical examples. Taxes on wage and salary incomes, by lowering take-home pay, tend to discourage some potential workers, who either withdraw from the labor force or work fewer hours than they otherwise would. These labor supply effects are most pronounced among those women, elderly, and young workers whose commitment to full-time employment is not ironclad. At the same time, of course, payroll taxes make workers more expensive to employers and can induce employers to seek cheaper alternatives, such as moving operations overseas or automating them.

The greater the responsiveness of labor supply to changes in after-tax wage rates, the greater the economic burden of income and payroll taxes. The economic loss involved when labor supply is reduced is the difference between the productive value of that forgone work and its opportunity cost to the employee. The value of labor to the employer is at least as great as her wage bill (inclusive of taxes and benefits), or she'd start laying off workers. The opportunity cost of work to the employed is less than his after-tax earnings, or he'd quit. Therefore, the economic cost of the lost labor input can be measured by the difference between its before-tax and after-tax cost; that is, by the marginal tax rate. The economic loss is higher, the higher the tax rate.

Taxes on income from investments have analogous economic costs. They lower the after-tax returns from investments and thereby induce people to seek tax-privileged investments or to save less. An influx of investments into tax shelters has an economic cost because capital is withdrawn from other investments that have a higher before-tax rate of return. When equilibrium is restored in capital markets, the after-tax (and risk-equivalent) rates of return on different investments will be equalized. Therefore, the before-tax returns across investments will differ by the differential tax rates. These differentials are a good measure, therefore, of the economic losses implied by the shift in each investment dollar toward less heavily taxed opportunities.

A lower rate of savings, by reducing capital formation, has long-lasting and powerful effects on economic productivity and growth. The more sensitive the savings rate is to the after-tax return on investments, the greater the economic cost of taxes on capital. As world capital markets become more highly integrated, the possibility grows that an increase in U.S. taxes on investment income could send U.S. savings abroad or reduce foreign investment in this country. As in the case of taxes on labor earnings, the economic loss involved is the loss of savings in response to a tax increase, valued at the marginal tax rate on savings.

Estimating these tax burdens is complicated. When the underlying issue is substituting environmental taxes for conventional income and profits taxes, the relevant measure is the gain from marginally reducing income and profits taxes when the revenues are made up by higher environmental taxes. Therefore, we can assume the level of government spending to remain constant. Moreover, abstracting from interpersonal changes in the distribution of tax incidence, which can be minimized through careful design of the tax package, we can assume that people's after-tax incomes are unchanged. Under these assumptions, the problem becomes one of estimating what is technically called the *marginal excess burden* of taxes.[1]

The appropriate measure of the sensitivity of labor supply to the after-tax wage is called the *compensated labor supply elasticity.*[2] There have been many attempts to estimate such supply elasticities

---

[1]There is a large variety of literature on the excess burden of taxation. A basic reference explaining the concept and estimating the excess burden of taxes on labor income is Browning (1987). See also Stuart (1984).

[2]This is the percentage response of hours worked to a small percentage change in after-tax hourly earnings, adjusted to eliminate the (usually negative) effects of higher incomes on hours worked.

for American male and female workers, using both actual labor market behavior and the results of income maintenance experiments. However, the numerical estimates still vary widely according to the data sources, analytic models, and econometric techniques used in the estimation process. Table 3.4.1 summarizes the findings of numerous studies (Pencavel 1986; Killingsworth and Heckman 1986).

The conclusion is that compensated labor supply elasticity for men is about 0.10 to 0.15. This means that a 10% rise in after-tax hourly earnings would induce a 1 or 1.5% rise in hours worked if after-tax income was kept constant. For women, the overall elasticity is higher, in the range of 0.3 to 0.5. Combining these estimates, using the relative shares of men and women in total labor hours worked in the United States as weights, gives an overall figure in the range of 0.25 to 0.30. A 10% rise in average hourly earnings would increase the labor supply by roughly 2.5%, other things equal.

The problems in estimating the responsiveness of savings to changes in the interest rate that savers can earn are considerably more complex, and researchers have produced an even wider range of estimates, ranging from nearly zero to one or more.[3] Most studies of the excess burden of taxes on capital income have used an aggregate estimate of around 0.4, while recognizing that the marginal tax rates on different forms of investment and savings differ substantially. (See Boskin 1978; Ballard et al. 1985; Fullerton and Henderson 1989; Jorgenson and Yun 1990; Trostel 1991).

These elasticity estimates have been used, along with measures of marginal tax rates, in several studies to determine the marginal excess burdens of taxes in the United States. All these studies come to two general conclusions: The burdens are high, and the burdens

[3]For a recent review, see Joel Slemrod, ed. (1990)

Table 3.4.1.[a] Supply of labor elasticity estimates

	Number	Range	Median	Mean
Male	39	0 to 0.84	0.1	0.11
Female	111	0 to 2	0.29	0.57

[a]Some data adjustment was made on outliers. For instance, since compensated labor supply should not be negative, negative estimates were not included. However, instead of discarding those low estimates or equally extreme estimates on the high end (over 2), they were included as 0 if they were less than 0 and as 2 if greater than 2. Of the 150 estimates, 29, or 19%, were changed in this way.

of taxes on income from investments are higher than those of taxes on labor incomes.[4] Table 3.4.2 summarizes several studies that suggest that the marginal excess burdens of payroll taxes are about $0.30 to $0.50 for every extra dollar of tax revenues collected thereby; that the marginal excess burdens of individual income taxes are in a somewhat higher range, $0.40 to $0.60 per dollar of additional revenues collected; and that the marginal excess burdens of taxes on income from investments are still higher, in the range of $0.60 to $1.20. These burdens are the additional loss of private income due to reductions in work effort and investment on top of the direct tax payment. Since some of these estimates were made before the tax cuts of the 1980s, which have lowered marginal tax rates on many incomes, the lower ends of these ranges are probably more applicable today.

These figures imply that government revenue needs are met through taxes that are extremely costly to the United States economy in terms of lost work and savings. If considerable government revenues could be raised in nondistortionary ways — allowing reductions in income, payroll, and profits taxes — the real economic savings would be huge.

Similar opportunities are available in the nation's cities and states, except that the potential gains to local economies are even larger. Throughout the nation, most state and local governments are

[4]Taxes on goods and services, such as sales and excise taxes, also create analogous economic burdens by creating a difference between the value of the taxed item to the purchaser and its cost of production. The same elements principally determine the marginal excess burden: the elasticity of supply of the taxed item and its marginal tax rate. The marginal excess burden of broadly based sales taxes is generally estimated to be lower than that of taxes on labor and capital. See Jorgenson and Yun (1990).

Table 3.4.2. Marginal excess burden estimates

Tax	Number	Range	Median	Mean
Social Security Payroll[a]	2	$0.31–$0.48	$0.40	$0.40
Individual income[b]	2	$0.40–$0.60	$0.50	$0.50
Investment income[c]	3	$0.58–$1.18	$0.92	$0.88

[a]Ballard (1991); Jorgenson and Yun (1990).
[b]Ballard (1991); Jorgenson and Yun (1990).
[c]Jorgenson and Yun (1990); Ballard (1991); Trostel (1991).

under severe fiscal pressure. Many have been facing budgetary crises. Since the mid-1980s, states and local government expenditures have outpaced revenues, undermining fiscal balances. The prior recessionary period exacerbated the imbalance, eroding tax bases while increasing both the demand for services and the costs of providing them. During 1991, with a projected budget gap totaling $40 to $50 billion in all the states, drastic actions were required. In 29 states, governments were forced to cut expenditures by more than $7.5 billion, while enacting tax increases totaling $10.3 billion (National Governors' Association 1991). In 1992, 35 states cut budgeted expenditures by a total of $5.7 billion while raising $15 billion in new taxes (National Governors' Association 1992). Further tax measures appear to be on the horizon.

Most state governments have little flexibility in dealing with these fiscal pressures. Laws or constitutional provisions require balanced budgets, even in recessionary periods. Since states already rely heavily on sales taxes, opportunities to raise them further to offset declines in consumer spending are limited. During fiscal year 1991, over 40% of state revenue increases came from personal and corporate income taxes (Belsie 1990).

These tax increases are measures of desperation. For the state economy, the problem is not just that a higher state personal income tax will induce some workers to work less; it is also that the higher state tax will induce some other workers to work less *within the boundaries of the state*. Other things equal, states that impose high personal taxes on their citizens will discourage immigration and encourage emigration. Similarly, the problem for the governor of a state is not just that a higher state tax on investment income will discourage savings; the problem is that a higher state tax on investment income will discourage investment *within the boundaries of the state* and encourage savings to flow elsewhere.

Because labor is somewhat mobile and capital is much more mobile among states, state governments inevitably find themselves in tax competition. Often, this competition takes the form of special tax incentives to attract new businesses. Competition also involves the level of state and local taxes. Inevitably, such competition penalizes states that raise their tax rates relative to those in force in neighboring and competing states. There is overwhelming evidence that state tax differentials influence the interstate movement of both capital and labor.

In a study on state and local economic development policies by Timothy Bartik (1991), 59 empirical studies of the effects of state and local taxes on intermetropolitan or interstate shifts in employment and business investment were included. These studies vary significantly in how they measure tax rates, differentials, and changes; in how they measure changes in employment or business location; and in how they control for other relevant factors, such as the quality of public services and infrastructure. Accordingly, the results can be used only to establish the weight of evidence and plausible ranges of relevant elasticities.

For example, five studies estimated the responsiveness of state employment to state and local income tax levels. The results are presented in the first row of Table 3.4.3. All five studies found that state and local personal income taxes have substantial and statistically significant effects on employment growth within the state, clustering around an estimated elasticity of 0.39. This indicates that jobs shift substantially among localities in response to state and local taxes.[5]

Bartik also reviewed studies measuring the effects of state and local taxes on business investment and location decisions. The second row in Table 3.4.3 summarizes 10 such studies that yielded estimates of the elasticity of investment with respect to state and local business taxes. Most of these studies controlled for general regional growth differentials and for differences among states in the

---

[5]An earlier literature on labor mobility is assessed by Michael Greenwood (1975).

Table 3.4.3. Labor and capital supply elasticities

Effect	Number	Range	Median	Mean
Taxes on labor supply: elasticity of employment with respect to state and local individual income taxes	5[a]	−0.66− −0.13	−0.39	−0.38
Taxes on capital supply: elasticity of business location or investment with respect to state and local corporate income taxes	10[b]	−1.4− −0.07	−0.17	−0.38

[a]Wasylenko (1988); Quan and Beck (1987), Carroll and Wasylenko (1989), Munnell (1990), Luce (1990).
[b]McConnell and Schwab (1990), Woodward (1990), Bartik (1989), Bauer and Cromwell (1989), Deich (1989), Papke (1986, 1987, 1989), Schmenner et al. (1987), Bartik (1985), Garofalo and Malhotra (1983), and Hodge (1981).

level of public investment. Again, the weight of evidence supports the commonsense conclusion that higher state taxes discourage business investment within the state.

Therefore, for local and state economies, increases in conventional taxes are doubly burdensome. They discourage work and savings as federal taxes do and, *in addition*, lead to a flight of labor and capital outside the tax jurisdiction. Since labor and capital are more likely to move in response to a change in incentives than to withdraw altogether from the economy, the economic loss to the state economy from a rise in state taxes, per dollar of revenue collected, is likely to be far greater than the loss to the national economy per dollar of new federal taxes.

This, along with serious revenue deficits, explains the search for tax alternatives in state government offices all across the nation. For 1993, only 25% of the proposed tax increases came from personal and corporate income taxes. State governments seem far more willing to impose "sin" taxes, user fees, and environmental charges than the federal government is. For 1993, over half of the new tax revenues are to come from increases in alcohol and tobacco taxes, gasoline taxes and motor vehicle registration fees, and other user fees. States already impose a wide variety of charges and fees related to environmental programs.

Environmental charges or taxes are one of many incentive-based instruments of environmental policy.[6] If applied appropriately, they share with others the virtue of promoting cost-effective control of environmental problems. If there are many actors contributing to a common environmental problem — many firms burning fossil fuels and producing carbon dioxide, for example — and the cost of cutting back the offending activities differs among firms, then a unit charge on that activity will encourage each actor to cut back to the extent that his per-unit abatement cost is less than the amount of the charge. Firms that can cut back at little cost will do so; those that would face much higher costs will cut back less. In the end, the unit tax will set a ceiling on costs to which all firms will adjust, and the total amount of environmental control induced by the tax will be achieved at minimum cost.

The potential gains from improved cost-effectiveness in U.S. environmental regulation are very substantial. As of 1992, the total

[6]Others include deposit-refund systems, non-compliance fees or fines, and marketable permits for environmentally damaging activities (such as emissions). Some analysts would extend the class of incentive-based policies to include liability laws, labeling requirements, and other measures.

cost of administering and complying with environmental regulations in the United States is around $120 billion per year, more than 2% of annual gross domestic product (Carlin et al. 1992, 12–44). Numerous studies of specific control programs have shown that actual costs are at least twice as high as the costs that would be incurred if cleanup and control responsibilities were reallocated to achieve cost-effectiveness (see Tietenberg 1985; South Coast Air Quality Management District 1992).

Taxes and charges, like other environmental policy instruments, are mechanisms to deal with systematic failures in market incentives that arise when individual actors are not confronted with the full costs (or benefits) of their activities. Such incentive failures are characteristic of environmental problems because environmental resources, such as air and water, are used in common and not readily divisible into privately owned parcels. When such resources are impaired—through the discharge of effluents, for example—the costs are diffused among all users. Unless incentive-based policies are in force, such costs cannot readily be collected from, or charged to, those whose activities result in damage. Consequently, environmentally damaging activities tend to be carried to excess.

Under some conditions, an environmental tax can induce not only a cost-effective distribution of control activities but also an overall level of control that minimizes the sum of environmental damages and control costs. The tax rate would have to be set to equal the marginal damages from an additional unit of the offending activity at just that overall level of control at which the marginal damage from an additional unit equals the marginal cost of abating it.

This situation is depicted in figure 3.4.1. The horizontal axis represents the level of the damaging activity, and the vertical axis the costs. The line *dd* portrays the additional private benefits that the actor derives from successive increments, assumed to decline, of the damaging activity. The line *cc* portrays the incremental private costs the actor incurs in increasing the level of activity. Taking only these private benefits and costs into account, the actor will choose a level of activity near the point *x*, which maximizes the private benefits net of costs. However, if the activity also imposes costs on others—by degrading an environmental resource that is used in common, for example—the total incremental costs as the activity expands might be portrayed as *c'c'*. The difference between the two cost curves represents what are called *external costs*, those not borne by the actor. At the level *x*, the activity results in incremental costs (to all

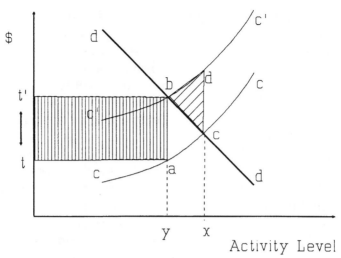

Figure 3.4.1. Marginal damage and marginal cost curves.

parties together) that are greater than the incremental benefits. These net losses are represented by the line *cd*. A level of activity that maximizes overall net benefits would be at the point *y*, at which marginal private benefits equals marginal private and external costs. So long as activity is above this level, each unit incurs net losses. The total loss is represented by the entire shaded triangle. A unit tax on the activity at a rate *tt′* would induce the actor to reduce the level of activity from *x* to *y*. The tax would bring in revenues in the amount of the rectangle *t′tab*, the tax rate times the tax base, which is the level of the activity after adjustment.

What is important to note in this simplified example is that, other than the costs of enforcement and administration, this tax does not create any excess burden. It has disincentive effects, but the activity that is discouraged is one that otherwise would be carried to excess and would cost society more at the margin than it is worth. In fact, when the level of the environmentally damaging activity is reduced from *x* to *y*, the tax results in economic savings amounting to the area of the shaded triangle. At each level of activity between *x* and *y*, the incremental private and external costs exceed the incremental benefits, resulting in losses. Avoiding those losses is a net saving to the economy. Thus, unlike taxes that discourage economically beneficial activities, such as work and savings, environmental taxes can discourage activities that, at the margin, cause economic losses. Therefore, rather than impose excess burdens, environmental taxes can provide revenues and economic gains.[7]

There is an enormous theoretical literature on environmental taxes and other incentive-based policy instruments.[8] The upshot is that taxes and charges are appropriate policy instruments when applied to certain kinds of environmental problems and are inappropriate—or inferior to other approaches when applied to others. The circumstances in which environmental charges are particularly appropriate include the following:

- The environmental problem is caused by the activities of numerous heterogeneous parties, so that private negotiations, permit trading, legal proceedings, or direct regulations would be difficult.
- Each party's actions contribute more or less proportionately, unit for unit, to the overall problem, so that particular "hot spots" or "bad actors" are not significant.
- The overall damages resulting from the activity are reasonably well understood and well behaved in that neither catastrophic damage thresholds nor rapidly decreasing marginal damage thresholds are likely to be encountered as the level of the activity increases.
- The parties differ significantly in characteristics that determine their costs of abating the problem, such as technology, age of equipment, availability of alternatives, scale, and so on.
- The dynamics of the problem are changing, so that a regulatory solution would soon be obsolete.
- The relevant behavior of each party can be monitored accurately at reasonable cost, so that incentive-based mechanisms linked to the level of the activity are enforceable.

---

[7]It has been pointed out that, if the revenues from environmental taxes can be used to reduce other distortionary taxes, then the environmental tax rate should be set to reflect these potential gains, as well as those reducing the environmental externality. The resulting tax rate could be either higher or lower than the rate tt depicted in figure 3.4.1, depending on the response of tax receipts to the tax rate at the rate tt (see Lee and Misiolek 1986). As a practical matter, this insight underscores the point that, under these conditions, the most appropriate rate of environmental taxes, where they are feasible, cannot be zero since a zero tax rate brings in no revenue. (See also Terkla 1984.))

[8]A classic treatment of the subject is Baumol and Oates (1975). A work focused more on problems of application is Anderson et al. (1977). A broader review of incentive-based policies discussing potential applications is contained in *Project 88* (1988) and *Project 88—Round II* (1991)

· A conflicting regulatory framework based on permitted technologies or emissions levels is not already functioning, so that
difficult problems of transition are not important considerations.

If these conditions are not met, then other incentive-based policies can probably achieve comparable or superior gains in efficiency
with less administrative difficulty. For example, a system of tradable
permits is typically easier to graft onto a command-and-control
(CAC) regulatory system based on permitted levels of emissions or
other environmentally damaging activities, provided that there aren't
too many regulated parties (see Tietenberg 1985). At least in principle, tradable permits also give regulatory agencies more control
over the level of the damaging activity, which is comforting if the
potential damages could escalate rapidly as the activity increased.

Nonetheless, many environmental problems that would not
easily yield to other incentive-based policies meet these conditions.
The discussion below deals in some detail with three of these:
municipal solid waste, urban highway congestion, and energy-
related pollution.

SOLVING THE SOLID WASTE PROBLEM

Between 1960 and 1988, the volume of annual municipal solid waste
generated more than doubled, from 88 million to 180 million tons.
The pace of new landfill construction has slowed as environmental
standards and community resistance have toughened. As a result,
landfill disposal costs can be expected to increase, making landfill
disposal only one, and a less attractive, option compared with others.

Since households in most communities pay for solid waste services
through property taxes, it costs them virtually nothing to put out an
extra trash bag, though the costs to communities in heavily urbanized
regions of dealing with more solid waste may be $100 per ton or more.
As a result, households do too little to cut back on waste disposal, and
communities are forced to spend too much on waste services.

This incentive problem can be corrected by charging households
the full incremental costs of waste disposal through a "pay-by-the-
bag" system. Hundreds of communities throughout the country have
now initiated such systems. According to our estimates, derived
from a statistical evaluation of the experiences of a sample of
communities, households that pay by the bag respond vigorously to
price signals. A typical community that raised its collection fee per

Table 3.4.4. Results of solid waste pay-by-the-bag systems

Based on market and nonmarket waste disposal costs	Communities	
	High-cost	**Moderate-cost**
Appropriate level of charges:		
per 32-gallon container ($)	1.83	1.03
per ton ($)	195	110
Changes in waste volumes:		
Reduction in landfill volume (lb/person/yr)	320	180
Increase in recycled volume (lb/person/yr)	133	75
For a community of 500,000 people:		
Reduction in landfill volume (%)	37	21
Net saving from landfill reduction ($ million/yr)	6.96	2.21
Increase in recycled volume (ton/yr)	29,688	16,741
Gross cost of recyling ($ million/yr)	2.97	1.67
Revenues from charges ($ million/yr)	23.57	16.73
For all high- and moderate-cost states:		
Net savings ($ million/yr)	487	618
Gross cost of recycling ($ million/yr)	206	467
Revenues ($ million/yr)	1,650	4,675

Source: Repetto et al. (1992), p. 28.

32-pound bag from zero to $1.50 — in line with incremental costs — would reduce solid waste generation by 18%. Fees combined with a curbside recycling program would reduce waste volume by about 30%. For pay-by-the-bag systems that include curbside recycling, the savings from reduced landfill costs would more than offset the budgetary costs of the recycling programs. Adopted nationwide in communities with high or moderate waste disposal costs, charges accompanied by curbside recycling would generate revenues of $6.3 billion per year and net savings of $432 million (see Table 3.4.4). This shows how local governments can use fee-for-service user charges to increase efficiency and take pressure off property taxes.

ABATING URBAN TRAFFIC CONGESTION

Variable tolls can help control traffic congestion on the nation's roads and highways. Congestion, which already costs tens of billions of dollars in delays, accidents, and pollution, has worsened because the total miles traveled by motor vehicles increased by 90% between 1970 and 1989 although funds were available to increase urban road capacity by less than 4%. Nearly 70% of rush-hour traffic endures stop-and-go conditions, a 30% increase in the past decade. Without a change in policies, congestion will only get worse.

Drivers ignore the full costs of using crowded roads. When drivers enter a congested highway, they consider only how long it will take to reach their destinations, ignoring the fact that their cars delay all other drivers and increase the probability of accidents. Therefore, too many cars are on the road under congested conditions. Tolls based on the costs that an additional car imposes on all others during congested periods would allocate road capacity more efficiently, inducing some drivers to reschedule or reroute trips, others to carpool, and still others to use public transportation.

The World Resource Institute's analysis estimates that tolls set to reflect the costs of traffic delays would range from zero to $2.10 for a typical 10-mile trip and reduce vehicle miles traveled at peak periods by 11% on the nation's busiest urban highways. Such tolls for the nation's urban areas would generate annual revenues of $44 billion and net savings exceeding $4 billion in reduced travel time, over and above the costs to drivers of adjusting their travel schedule. To cover the full social costs of accidents and delays, tolls would range from zero to about $3.60 for a typical 10-mile trip, would reduce peak traffic by 22%, and would save $11 billion per year on revenues totaling $98 billion (see Table 3.4.5). If congestion tolls are not adopted, nearly $50 billion will have to be spent on highway construction by 1999 just to achieve the same mitigating effect on road congestion. Congestion tolls could avoid these costs while generating the billions of dollars needed to pay for upkeep of our existing transportation infrastructure and to improve public transportation options. Peak-period pricing works for electricity and telecommunications, two other capital-intensive industries. It can work for urban transportation.

Fortunately, the technologies needed to make congestion tolls cheap and efficient already exist. Along several U.S. highways, electronic toll stations transmit signals to "smart cards" carried on passing vehicles—cards that record a vehicle's point of entry to and exit from the highway. Such systems are already in place on several U.S. highways. With congestion tolls, the card can be debited directly at rates depending on trip length, choice of route, and time of travel.

Table 3.4.5. Results of a nationwide congestion toll system, 1989

	Congestion toll
Original VMT[a]	1,055,637 (million annually)
Adjusted VMT	989,153 (million annually)
Percent reduction	6.3
Toll range	0–21 (cents/mile)
Revenue generated	44.1 (billion dollars annually)
Most congested VMT	399,432 (million annually)
After toll	354,964 (million annually)
Percent reduction	11.1
Welfare gains	4.2 (billion dollars annually)
	With accident toll
Adjusted VMT	966,708 (million annually)
Percent reduction	8.4
Toll range	0–28 (cents/mile)
Revenue generated	73.4 (billion dollars annually)
Adjusted congested VMT	332,971 (million annually)
Percent reduction	16.6
Welfare gains	7.3 (billion dollars annually)
	With accident delay toll
Adjusted VMT	943,912 (million annually)
Percent reduction	10.6
Toll range	0–36 (cents/mile)
Revenue generated	98.4 (billion dollars annually)
Adjusted congested VMT	310,455 (million annually)
Percent reduction	22.3
Welfare gains	10.8 (billion dollars annually)

Source: Repetto et al. (1992), p. 45.
[a](VMT: Vehicle mile traveled)

REDUCING ENVIRONMENTAL IMPACTS
FROM ENERGY USE

Throughout the fuel cycle, there are significant environmental damages. At the extraction stage, there are problems with land disturbance, mine drainage and wastes, oil spills, ecological disruptions from hydroelectric storage, and so on. At the conversion stage, there are impacts on land, air, and water quality. In 1992, atmospheric emissions in the United States totaled 20 million tons of SO_x, 19 million tons of NO_x, 62 million tons of CO, 17 million tons of VOC, and 7.5 million tons of particulates. The large majority of these emissions, most of which have been partially reduced, emanate from energy use in transportation, electricity generation, and industrial processes.

Many of these impacts are addressed by environmental regulations, with varying degrees of effectiveness. Nonetheless, there are still significant environmental damages associated with energy conversion and use that are not captured by market prices. For example, recent estimates of the environmental damages due to atmospheric emissions from coal-fired power stations, at current standards of pollution control, equal at least \$0.006/kWh, approximately 10% of total generating costs (U.S. EPA 1992).

Combustion of fossil fuels also generates carbon dioxide which, along with other greenhouse gases, threatens to bring about substantial warming of the Earth's atmosphere. Unless checked by effective national and international policy, carbon dioxide emissions will continue to grow in the United States and worldwide. Global warming could cause significant environmental damage, including coastal erosion and flooding from sea level rise, the destruction of wetlands and other ecosystems, accelerated species extinction, and disruption of hydrological patterns.

Stabilizing atmospheric concentrations of greenhouse gases requires reducing CO_2 emissions. Because almost all economic activities use energy derived from fossil fuels, such reductions could be achieved most efficiently by taxing the carbon content of fuels. Because the carbon content of fuels varies per unit of energy, coal, oil, and natural gas would be taxed at different rates. A carbon tax would provide market incentives for all users to find the best mix of fossil and nonfossil fuels and energy conservation for their particular circumstances, avoiding the inefficiencies of regulatory mandates.

A carbon tax of about \$30 per ton phased in over five years would stabilize U.S. emissions at 1990 levels by the year 2000, generating revenues of $36 billion by the fifth year. Most macroeconomic models suggest that the economic consequences of such a tax would be either fairly small losses or outright gains, depending on how the tax revenues were recycled into the economy through other tax cuts. But these macroeconomic models neglect the potential damages from climate change and also overlook other significant benefits of carbon taxes — reduced dependence on oil imports and decreased emissions of other air pollutants.

Since the main impact of a carbon tax would be to reduce the growth of coal consumption, measures would be needed to offset the impacts on such states as West Virginia, Kentucky, and Wyoming, where coal production is concentrated. An alternative is a modified Btu tax, which taxes fuels in proportion to their Btu content or Btu equivalent, with disproportionately heavy rates on oil. While less efficient in terms of CO_2 abatement than a strict carbon tax, the Btu tax deals with a broader range of environmental and security issues and is estimated to have similar impacts on CO_2 emissions, per dollar of revenue, over a 15-year horizon.

The economic burden of energy taxes would be small. Energy prices in the United States continue to be below those in Europe and Japan, our principal industrial competitors. Studies show little competitive advantage from low energy prices. Countries with low energy prices have not experienced more rapid economic growth, lower rates of inflation, or more favorable trade balances. Principally, low energy prices lead to continuing low levels of energy efficiency. Energy costs make up only 2.6% of U.S. industrial production costs, on average, and, even if totally absorbed by industrial energy users, the tax would therefore raise average manufacturing costs by 0.01%.

Moreover, since most energy-intensive industries are also capital-intensive industries, the indirect economic effects on industrial competitiveness will be favorable. The U.S. current account deficit has been largely a monetary phenomenon, arising out of the need for huge foreign borrowing to finance domestic deficits. Reducing the budgetary deficit will reduce imports and long-term interest rates and will prevent the dollar from rising against other currencies. Capital-intensive industries might gain more from lower interest rates and a more favorable exchange rate than they lose from higher energy prices. Bond and foreign exchange markets have reacted precisely in this direction in anticipation of enactment of the Administration's deficit-reducing proposals.

Also important is the effect on employment. An energy tax falls most heavily on the most energy-intensive industries, which are not, by and large, the most labor-intensive. Morever, the energy tax encourages all firms to substitute other production inputs for energy use. This generally implies a substitution toward increased employment. Given the concern over the slow rate of job growth in the current recovery, this distinction is important and favors an energy tax over such alternatives as payroll taxes or a value-added tax.

The distributional effects of a broad-based energy tax are mildly regressive as a percentage of household expenditures. So are payroll and value-added taxes. The principal difference is that an energy tax offers easy and constructive opportunities for tax savings. At present, the only way most people can reduce their tax bills is to work less and earn less. Environmental charges would give them the option of reducing their tax bills by acting on their principles — for instance, saving fuel by bicycling to work.

OTHER POSSIBLE GREEN FEES

There is a wide range of other potentially useful environmental charges, including effluent charges on toxic substances and vehicle emissions, recreation fees for use of the national forests and other public lands, product charges on ozone-depleting substances and agricultural chemicals, and the reduction of subsidies for mineral extraction and other commodities produced on public lands. Such environmental charges would reduce a great many damaging activities in a cost-effective way while raising at least $12 billion in revenues (see Table 3.4.6).

Thus, according to our estimates, switching some of the revenue burden from distortive taxes on income, employment, and profits to environmental taxes on resource waste, collection, and pollution would yield double economic benefits. A reduction in tax rates on income and profits would reduce the marginal excess burden by $0.40 to $0.60 per dollar of reduced tax revenue. If those revenues were regained through environmental taxes, the *additional* net economic savings would range from $0.05 to $0.20 per dollar of revenue. These additional savings are the benefits from averted environmental damages net of the incremental costs of environmental protection. Putting these parts together yields the striking conclusion that the total possible gain could easily be $0.45 to $0.80 per dollar of tax shifted from "goods" to "bads," with no loss of revenues. The gains would come in the form of improved environmental quality, reduced infrastructure needs, higher rates of savings and investment, increased employment, and faster productivity growth.

Table 3.4.6. Illustrative options for environmental charges, by category

I. Effluent or emissions charges:
  1. On water effluents permitted under the National Pollutant Discharge Elimination System (NPDES)
  2. On toxic releases documented in Toxics Release Inventory
  3. On vehicular emissions in clean air nonattainment areas
  4. Solid waste collection and disposal charges

II. Charges on environmentally damaging activities:
  1. Recreational user fees on public lands
  2. Highway congestion tolls
  3. Noise charges on airport landings
  4. Impact fees on installation of septic systems, underground storage tanks, construction projects with environmental impacts, etc.

III. Product charges:
  1. Taxes based on the carbon content of fossil fuels
  2. Gasoline taxes
  3. Excise taxes on ozone-depleting substances
  4. Taxes on agricultural chemicals
  5. Taxes on virgin materials

IV. Deposit-return charges:
  1. On vehicles
  2. On lead-acid and nickel-cadmium batteries
  3. On vehicle tires
  4. On beverage containers
  5. On lubricating oil

V. Reduction of tax benefits and subsidies:
  1. Percentage depletion allowances for energy and other minerals
  2. Percentage depletion allowances for groundwater extraction
  3. Charging market royalties for hard-rock mining on public lands
  4. Eliminating below-cost timber sales
  5. Charging market rates for grazing rights on public lands
  6. Charging market rates for state and federal irrigation water
  7. Charging market rates for federal power

Source: Repetto et al. (1992), p. 73.

The environmental charges discussed above could yield at least $100 billion in annual revenues, split among federal, state, and local governments. Congestion tolls on urban highways could generate $40 to $90 billion, energy taxes $20 to $30 billion, solid waste charges another $5 to $10 billion, and other environmental fees and charges $10 to $12 billion. Using these revenue sources would allow governments to reduce marginal rates of distortionary taxation substantially and produce $45 to $80 billion in annual net economic benefits.

Of course, there is no reason why *all* the revenues from such environmental charges need be recycled through reductions in other, more distorting taxes. Some might be used for specific expenditure programs to compensate citizens who are disproportionately hit by environmental charges or to make the charges more effective. Spending some of the money from congestion tolls on public transport and spending some revenues from solid waste charges on community recycling programs are illustrations of these options. Some of the revenues might be used for deficit reduction.

If the economic trade-offs from such tax shifts are so favorable, what about the political trade-offs? Would such a shift in the tax base be politically acceptable? Would it be fair? The answer undoubtedly depends on the way the issue is framed. If people are asked whether they favor higher taxes, the answer is overwhelmingly no. If people are asked whether they would rather be taxed on their use of energy and on the amount of waste they generate than on their salaries, profits, or monthly expenditures, the answer is yes. According to public opinion polls, most people faced with a choice of higher taxes would prefer "sin" taxes on cigarettes and alcohol, or pollution taxes, because they see some direct benefit coming from those tax payments.

These findings refute the argument that environmental quality can be obtained only at the cost of lost jobs and income. Indeed, providing a better framework of market incentives by restructuring our revenue system can simultaneously improve environmental quality and make the American economy much more competitive. Taxes on income, payrolls, profits, and value added are distortive taxes, and their use implies some net welfare costs to the economy. By contrast, well-designed energy and environmental taxes are corrective taxes and can achieve welfare gains by reducing excessive environmental damages at the same time that they raise government revenues.

REFERENCES

Anderson, Frederick R., Alan V. Kneese, Phillip D. Reed, Russell Stevens, and Serge Taylor. 1977. *Environmental Improvement Through Economic Incentives*. Baltimore: Johns Hopkins University Press for Resources for the Future.

Ballard, Charles L. 1991. "Marginal Efficiency Cost Calculations for Different Types of Government Expenditure: A Review." Paper presented at the Australian Conference in Applied General Equilibrium, Melbourne, Australia. May 27–28.

Ballard, Charles L., John B. Shoven, and John Whalley. 1985. "General Equilibrium Computations of the Marginal Welfare Costs of Taxation in the United States." *American Economic Review* 106(1):128–138. March.

Bartik, Timothy J. 1991. *Who Benefits from State and Local Economic Development Policies?* Kalamazoo, MI: W. E. Upjohn Institute for Employment Research.

Bartik, Timothy J. 1989. "Small Business Start-ups in the United States: Estimates of the Effects of Characteristics of States." *Southern Economic Journal* 55(4):1004–1018.

Bartik, Timothy J. 1985. "Business Location Decisions in the United States: Estimates of the Effects of Unionization, Taxes, and Other Characteristics of States." *Journal of Business and Economic Statistics* 3(1):14–22.

Bauer, Paul W., and Brian A. Cromwell. 1989. "The Effect of Bank Structure and Profitability on Firm Openings." *Economic Review* 25(4):29–39.

Baumol, William J., and Wallace E. Oates. 1975. *The Theory of Environmental Policy*. Englewood Cliffs, NJ: Prentice-Hall.

Belsie, Laurent. 1990. "Budget Cuts Will Squeeze Local Treasuries." *Christian Science Monitor*. October 15.

Boskin, Michael J. 1978. "Savings, Taxation and the Rate of Interest." *Journal of Political Economy* 86 (2, II):3–28.

Browning, E. K. "On the Marginal Welfare Cost of Taxation." *American Economic Review* 77(1):11–23. March.

Carlin, Alan, Paul F. Scodari, and Don H. Garner. 1992. "Environmental Investments: The Costs of Cleaning Up." *Environment* 34(2). March.

Carroll, Robert, and Michael Wasylenko. 1989. "The Shifting Fate of Fiscal Variables and Their Effect on Economic Development." In *Proceedings of the Eighty-Second Annual Conference on Taxation*. Columbus, OH: National Tax Association—Tax Institute of America. October 8–11. Pp. 283–290.

Deich, Michael. 1989. "State Taxes and Manufacturing Plant Location." In *Proceedings of the Eighty-Second Annual Conference on Taxation*. Columbus, OH: National Tax Association—Tax Institute of America. October 8–11.

Fullerton, Don, and Yolanda K. Henderson. 1989. "The Marginal Excess Burden of Different Capital Tax Instruments." *Review of Economics and Statistics* 71(3):435–442. August.

Garofalo, Gasper A., and Devinder M. Malhotra. 1983. "Regional Capital Formation in U.S. Manufacturing During the 1970s." *Journal of Regional Science* 27(3):391–401. August.

Greenwood, Michael. 1975. "Research on Internal Migration in the United States: A Survey." *Journal of Economic Literature* 13:91–112. June.

Hodge, James H. 1981. "A Study of Regional Investment Decisions." In *Research in Urban Economics* 1, edited by J. Vernon Henderson. Greenwich, CT: JAI Press.

Jorgenson, Dale, and Kin-young Yun. 1990. "The Excess Burden of Taxation in the United States." Mimeo. Cambridge, MA: Harvard Institute for Economic Research. November.

Killingsworth, M. R., and J. J. Heckman. 1986. "Female Labour Supply." In *Handbook of Labor Economics*, Vol. I, edited by Ashenfelter, Orley, and Richard Layard. New York: Elsevier Science Publishers, pp. 103–200.

Lee, Dwight R., and Walter S. Misiolek. 1986. "Substituting Pollution Taxation for General Taxation." *Journal of Environmental Economics and Management* 13(4):333–354. December.

Luce, Thomas F. Jr. 1990. "The Determinants of Metropolitan Area Growth Disparities in High-Technology and Low-Technology Industries." Working Paper. Department of Public Administration, Pennsylvania State University, University Park.

McConnell, Virginia D., and Robert M. Schwab. 1990. "The Impact of Environmental Regulation on Industry Location Decisions: The Motor Vehicle Industry." *Land Economics* 66(1):67–81.

Munnell, Alicia H. 1990. "How Does Public Infrastructure Affect Regional Economic Performance?" *New England Economic Review* (September/October): 11–33.

National Governors' Association. 1992. Fiscal Survey of the States. Washington, DC: National Governors' Association. April.

National Governors' Association. 1991. Fiscal Survey of the States. Washington, DC: National Governors' Association. October.

Papke, Leslie E. 1989. "Interstate Business Tax Differentials and New Firm Location: Evidence from Panel Data." National Bureau of Economic Research Working Paper 3184. November.

Papke, Leslie E. 1987. "Subnational Taxation and Capital Mobility: Estimates of Tax-Price Elasticities." *National Tax Journal* 40(2):191–204. June.

Papke, Leslie E. 1986. "The Location of New Manufacturing Plants and State Business Taxes: Evidence From Panel Data." In *Proceedings of the Seventy-Ninth Annual Conference on Taxation.* Hartford, CT: National Tax Association-Tax Institute of America. November 9–12.

Pencavel, John. 1986. "Labor Supply of Men: A Survey." In *Handbook of Labor Economics*, Vol. I, edited by Ashenfelter, Orley, and Richard Layard. New York: Elsevier Science Publishers, pp. 5–102.

Repetto, Robert C., Roger Dower, Robin Jenkins, and Jacqueline Geoghegan. 1992. "Green Fees: How a Tax Shift can Work for the Economy and the Environment." World Resources Report. Washington, DC: World Resources Institute. November.

Quan, Nguyen T., and John H. Beck. 1987. "Public Education Expenditures and State Economic Growth: Northeast and Sunbelt Regions." *Southern Economic Journal* 54(2):361–376. October.

Schmenner, Roger W., Joel C. Huber, and Randall L. Cook. 1987. "Geographic Differences and the Location of New Manufacturing Facilities." *Journal of Urban Economics* 21:83–104.

Slemrod, Joel (ed.). 1990. *Do Taxes Matter? Lessons from the 1980s.* Cambridge, MA: MIT Press.

South Coast Air Quality Management District. 1992. *RECLAIM: Regional Clean Air Incentives Market.* Los Angeles, CA: SCAQMD.

Stuart, Charles. 1984. "Welfare Costs per Dollar of Additional Tax Revenue in the United States." *American Economic Review* 74(2):352–362. June.

Terkla, David. 1984. "The Efficiency Value of Effluent Tax Revenues." *Journal of Environmental Economics and Management* 11(4):107–123. December.

Tietenberg, Thomas H. 1985. *Emissions Trading: An Exercise in Reforming Pollution Policy.* Washington, DC: Johns Hopkins Press for Resources for the Future, p. 65.

Trostel, Phillip A. 1991. "Taxation in a Dynamic General Equilibrium Model With Human Capital." Ph.D. Dissertation. Department of Economics, Texas A&M University, College Station.

U.S. EPA. 1992. *Renewable Energy Generation: An Assessment of Pollution Prevention Potential.* Washington, DC: U.S. Environmental Protection Agency.

Wasylenko, Michael J. 1988. "Economic Development in Nebraska." Nebraska Comprehensive Tax Study Staff Paper No. 1, rev. Metropolitan Studies Program, the Maxwell School, Syracuse University, Syracuse, NY.

Wirth, Timothy E. and John Heinz. 1988. *Project 88. Harnessing Market Forces to Protect Our Environment.* Washington, DC: Senate Office Building. December.

Wirth, Timothy E. and John Heinz. 1991. *Project 88—Round II. Incentives for Action: Designing Market-Based Environmental Strategies.* Washington, DC: Senate Office Building. May.

Woodward, Douglas P. 1990. "Locational Determinants of Japanese Manufacturing Start-ups in the United States." Working Paper. University of South Carolina, Columbia.

3.41

❏

DISCUSSANT

Paul J. Pieper

Robert Repetto makes a strong case for using incentive systems in the design of tax and environmental policy. For economists, the importance of incentives is almost second nature, and so, for me, the paper was preaching to the converted. Judging from how policies are made, however, incentive systems are not second nature to everyone; and so I think there are many "heathens" out there who need to be converted. Therefore, even though much of the paper is relatively noncontroversial for economists, I think it is very valuable for its clear exposition of the economic rationale for green taxes and other incentive mechanisms. Murphy's law of economics states says that economists have the most influence in areas in which they agree least and the least influence in areas in which they agree most. Unfortunately, I think that, to date, the design of environmental incentive systems pretty much fits the second part of Murphy's law. I hope that this paper will receive wide circulation among policy-makers, so that incentive systems and green fees will have more influence than they currently do.

Since I agree with the main thrust of the paper, I will limit my comments to a few of the more peripheral issues. First, as a macro-economist, I would like to address the issue of jobs versus the environment, which I believe is a nonissue. In the aggregate, the number of jobs is determined by macroeconomic factors and not by environmental policies, which are more microeconomic in nature. Environmental policy may change the compositon of output and, hence, jobs but, as long as the labor market is well functioning, with reasonable mobility of labor and adjustment of wages to clear the market, it will not significantly affect the aggregate number of jobs. There will be winners and losers, but the jobs in the winning

284

industries will tend to offset those in the losing industries, at least after a period of adjustment. I would use as an analogy the issue of free trade which, it is sometimes argued, may have a negative effect on jobs. Loss of jobs from free trade may be highly visible, yet there are less visible job gains offsetting these losses.

I would also say, in passing, that I disagree with Howard Klee's comments, which follow, about the effect of an energy tax on net exports. Again, from a macroeconomic perspective, you cannot assume that the exchange rate will remain unchanged after the tax. If an energy tax did raise costs and put U.S. exporters at a disadvantage, then the dollar would fall, which would reduce or even eliminate the adverse effect on exports.

I think Repetto makes a good point in maintaining that a properly designed environmental tax that substitutes for taxes on labor and capital can have a double benefit. First, it can eliminate the negative environmental externality and, second, it can also reduce the distortive effect of taxes on labor and capital. However, the estimates of the marginal excess burden of labor and capital taxes shown in Table 3.4.2 seem quite high to me. I am not familiar with this literature, but my intuition is that the burden of these taxes is considerably less than $0.40 or $0.50 on the dollar. If the marginal burden were this high, one would expect to have seen more of an effect on output from the reduction in marginal taxes during the 1980s.

One general point I would like to make is that a higher level of taxes will increase the incentive for tax evasion. Take as an example the pay-per-bag garbage collection, which may lead to an increase in tax evasion in the form of illegal dumping. If this behavior is large enough, you will have a bigger problem than what you had under the fixed fee system. It is important to know whether this has occurred in areas that have instituted pay-per-bag fees; perhaps the fees now set are too low for illegal dumping to be significant. In any case, the possibility for evasive behavior must be taken into consideration in the design and implementation of tax policies.

My next point concerns using taxes to reduce congestion. I am a little unclear about how the high-tech system for collecting tolls using smart cards works. I assume that a machine reads your card as you drive by and you get the bill in the mail later. I have some question about enforcement. Do people treat the notices like Chicago parking tickets and just ignore them? I wonder how the system deals with out-of-state cars without cards. Again, although I am not familiar with the mechanics of these systems, I would expect that there would have to be at least one attendant on duty for each entrance to handle cars without cards. It would therefore appear to be expensive to operate such a system on many urban highways where there are exits every few hundred yards. An optimal policy must consider the cost of tax collection, and widespread use of smart-card systems may be very expensive.

A less expensive alternative would be to increase the gasoline tax, which would discourage driving by raising its marginal cost. Granted, the gasoline tax could not be used to charge differential prices at peak periods, as with the smart-card system, but it would be much less costly to collect. It would also discourage driving in all areas whereas, because of collection costs, the effect of the smart-card system would be limited to major highways. The smart-card system might also lead to increased congestion on lesser, toll-free arteries.

A second alternative is being used in Singapore, a very densely populated city. This system requires a special sticker to drive at peak periods. The supply of these stickers is limited to the optimal carrying capacity of the roads and then auctioned off. This is a rather draconian approach, but it illustrates one among a number of alternative approaches.

Finally, I have one minor comment concerning the carbon tax; the paper states that measures would be needed to offset the impact of the tax on coal-producing states such as West Virginia, Kentucky, and Wyoming. It think that, politically speaking, this statement is correct. Given that Senator Robert Byrd is from West Virginia, no carbon tax would pass Congress unless such measures were taken. But, economically speaking, it is not clear that this should be done. It may be better to create incentives for the labor released from these industries to migrate to other areas rather than trying to use subsidies to get industries to migrate to those states.

Let me close by reiterating that I am very much in sympathy with the paper's main message, that of using taxes and fees to help meet environmental goals, and I hope that the author continues his research in this area.

3.42

❑

DISCUSSANT

Howard Klee Jr.

I think Bob's paper has much to recommend it. I can't argue with the logic of encouraging activities that are good, like saving resources and improving the environment; and discouraging activities that are bad, like waste, traffic congestion, and inefficiency. However, I have some concerns about the underlying premise that the government needs more revenue.

Government functioned reasonably well in the '60s when it consumed about 19% of the gross domestic product. The budget was balanced, or nearly so, and revenue equaled spending. Today, government revenue is still about 19% of the gross domestic product, but spending is closer to 24%. I am not convinced that the government is functioning a lot better or that we are getting a lot more benefits for that extra 5% or 6% that is being spent. I just don't think it's appropriate to consider raising (green) taxes until the root cause of past regulatory mismanagement or excessive government growth has been defined and addressed.

Amoco recently completed a multiyear study with the U.S. EPA at our Yorktown, Virginia, refinery. We learned many things from this study: the value and benefits of government and industry partnerships, which are quite rare; the value of having sound data before we start making decisions; and the need to prioritize the problems that we're going to address. It think its particularly germane to today's discussion that we learned that we could achieve what I would describe as an equivalent level of environmental protection at this facility for about 20% of the cost of current and pending government-mandated programs. This puts me squarely in agreement with Bob's opening comments, that it is possible to have a strong economy and a healthy environment. But we need to be smarter buyers of environmental protection.

In my opinion, 80% of the money that we are spending in this facility for environmental protection is generally ineffective. We are not purchasing any additional protection for people, plants, or animals. No additional jobs are created, no additional product is produced, no additional revenue and no local income taxes are generated. In fact, our income tax obligation, at least theoretically, would decrease because of the depreciation for the equipment that we are required to install that is not providing any measurable benefits. Yorktown is a small refinery. By being more intelligent about how we achieve environmental quality, Amoco found we might save close to $40 million over the next 10 years at this small facility, all the while protecting the environment, the surrounding community, and the ecosystems there. Based on our work, I'm personally confident that there are similar savings possible in a variety of other states and manufacturing facilities.

I've seen and heard a lot about the desired effects of a Btu tax. Bob's paper discusses several of these, but I think that relatively little has been said about the possible undesirable consequences. A few deserve some further consideration. With a Btu tax, U.S. businesses will be less competitive internationally. Energy prices will rise and discriminate against U.S. manufactured goods that have significant energy components. For example, aluminum producers in other countries that are not paying an energy tax on Btu's consumed will have a much lower cost.

Seeking to expand exports, which have been one of the bright spots in the U.S. economy, does not match up with increasing their costs. Some people argue that other countries have much higher energy prices already and, therefore, we're just catching up with them. It is true that *consumer* energy prices are higher, but that is generally *not* true at the manufacturing levels.

3.5

❏

JOINT AMOCO/EPA POLLUTION PREVENTION PROJECT

Ronald E. Schmitt

INTRODUCTION

❏ All of us want to breathe clean air, drink clean water, and enjoy our rich lands. We believe that the Yorktown pollution prevention project marks a new approach to achieving the goals of improved air, land, and water quality. Certainly, industry and government share these goals, but they are often at odds over ways of reaching them. In this project, the regulated and the regulator got involved cooperatively to evaluate pollution prevention at an operating facility. Before describing the project and its findings, let me provide you with some background to show why the project was initiated.

Background to the Project

At the time of the study, Amoco was composed of three operating companies. Amoco has since changed its organizational structure, but the objectives of these three business sectors remain the same as before the project. The Amoco exploration and Production Sector explores and produces crude oil throughout the world. The Amoco Chemical Sector produces commodity chemicals and consumer products at plants located throughout the world. Amoco Petroleum Products Sector operates five refineries in the United States, which process nearly a million barrels of crude oil a day, providing gasoline and other fuels, lube oils and other products. Unlike other oil companies, which operate refineries in other countries, Amoco refines and distributes products only in the United States. Amoco's survival depends on its ability to compete in the global market, and so it is keenly interested in ensuring that environmental benefits can be achieved in the United States without placing domestic companies at a disadvantage in relation to those operating in other countries.

Environmental legislation and regulation have grown in complexity, and compliance costs have skyrocketed. The concern, however, is not the amount of resources expended. Rather, the question is whether we are achieving genuine environmental improvements for the outlays. And are we doing the things we should be doing to protect the environment? From the viewpoints of both government and industry, it seemed there had to be a better approach to achieve the goal both sides wanted: a cleaner environment. Faced with those challenges and frustrations, Amoco and EPA, in late 1989, began talking about the possibility of cooperating in a pollution prevention project.

Pollution Prevention

At the time this project began, pollution prevention was a concept predicated on reducing or eliminating releases of materials into the environment, termed *source reduction*, rather than managing the releases later. An EPA–Amoco Oversight Workgroup adopted this general concept and agreed to consider all opportunities—source reduction, recycling, treatment, and environmentally sound disposal—as potential choices in pollution management. This project was an opportunity to test how the application of that definition differed across industrial facilities, especially extractive industries like petroleum refineries, whose feedstocks inherently contain wastes that cannot easily be eliminated by feed substitution.

Potential Risks and Benefits

Such an unprecedented project had many risks, including the possible discovery of noncompliance, waste of time and money (eventually totaling $2.3 million) if the study results were not used, and the appearance that the EPA was too close to an industry that it regulated. But the study also had many potential benefits. For example, each side would have the opportunity to become better

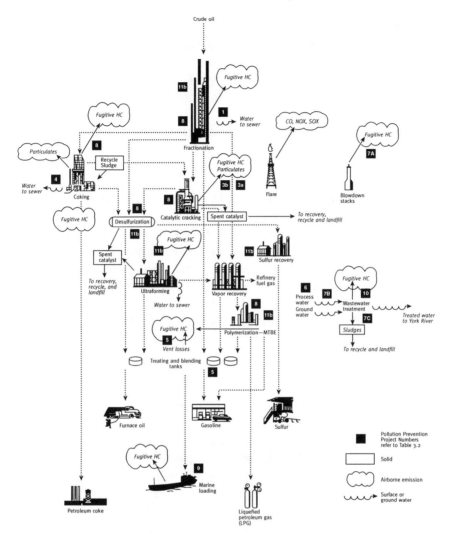

Figure 3.5.1. Yorktown refinery: simplified flow diagram with release sources and control options.

educated about the problems of its counterpart. At the same time, working side by side would give participants a chance to build better relationships, which might lead to improved drafting and administering of regulations.

More concretely, the Yorktown project provided for an overall, instead of piecemeal, examination of pollution prevention. In place of individual examinations of water, air, or land emissions, investigators would conduct an integrated review to discover what interaction takes place between these media.

This integrated review had the potential to point out opportunities for improving the methods employed to prevent pollution—or to identify those areas requiring additional research.

Moreover, the project would be carried out in an actual, operating industrial facility, providing a unique opportunity to generate new ideas for improving our environment. Never before had regulators and the regulated been able to work side by side in a real-life facility.

DECISION TO PROCEED AT YORKTOWN

Acknowledging the risks and anticipating the benefits, Amoco and EPA decided to go forward. There were no guarantees or promises exchanged on the use of information, such as immunity for violations discovered. Instead, both parties agreed to get involved on the basis of a simple verbal agreement. Mutual trust was no small part of this project. Both parties were working with previous adversaries, and the success of the project depended on each side placing its trust in the other.

Yorktown Refinery

Amoco's Yorktown, Virginia, petroleum refinery was chosen as the site for study. The facility, located on 1400 acres along the York River near Chesapeake Bay, is capable of processing 53,000 barrels of oil per day into gasoline, heating oil, and other products.

This size and product mix are representative of the petroleum industry, yet the Yorktown Refinery was small enough to permit a thorough study within the two-year period planned for the project. Also, its proximity to Washington, DC, and its location in an environmentally sensitive area made the refinery a logical choice. During the course of the project, over 100 EPA and Virginia regulatory professionals visited the refinery.

Project Organization

Over 200 people from various organizations, including Amoco, EPA, the Commonwealth of Virginia, and many others, contributed to the project. Dr. Howard Klee of Amoco and Dr. Mahesh Podar of the EPA codirected the project. A workgroup composed of Amoco and EPA representatives was formed to provide project oversight. Since this was a new and challenging project for both parties, the workgroup decided to limit outside participation in the project until Amoco and the EPA had a chance to learn to work together.

At a later stage, the project made use of outside experts who could provide peer review. The EPA chose Resources for the Future to select the people to serve on the committee. Twelve experts, representing many disciplines, were chosen. Associated with government agencies, private consulting firms, and universities, they analyzed the approach, methods, and findings, and provided independent, informed opinions on the validity of each step in the project.

Project Goals

The project workgroup developed a simple set of goals. They agreed that the project should:

1. Determine the types, amounts, and sources of emissions that the refinery releases to the air, land, and water.
2. Develop options to reduce these releases; determine the benefits, impacts, and costs of different options; and select the most cost-effective option for improving environmental quality.
3. Identify factors that encourage or discourage pollution prevention initiatives.
4. Increase participant's knowledge of refinery and regulatory systems.

Exclusions, Limitations, and Unique Features

A number of areas were specifically excluded or limited in this project. For example:

1. Sampling time and data provided a "snapshot" of releases rather than measured annual values.
2. Available technologies were considered rather than innovative techniques or those that would require research and development.
3. Chemical changes of airborne pollutants were not evaluated.
4. Data and analysis focused on the Yorktown Refinery. Broader regional concerns were not evaluated.
5. Emergency and upset events were not studied in this project.

Several aspects of this project made it unique and innovative:

1. The project was an entirely voluntary initiative between Amoco and the U.S. Environmental Protection Agency.
2. No contracts or written agreements were required to commit to work together on the study.
3. Extensive sampling and other data collection provided a "sound science" framework for the investigation.
4. Public perception was examined and incorporated into the alternative ranking and findings of the study.
5. A multidimensional decision analysis tool was utilized to weight factors determined to be important to the ranking of identified emissions reduction alternatives.

EXTENSIVE SAMPLING PROGRAM

The Yorktown Refinery had good information about the quantity of material released to the York River from the National Pollution Discharge Elimination System permit monitoring requirements and about solid wastes as a result of internal programs and participation in recent American Petroleum Institute surveys (API 1991). These releases, however, made up only 11% of the total releases from the facility. Available data did not include adequate chemical-specific characterization of the water discharge or solid waste streams.

For the following reasons, the refinery (and other refineries as well) could not easily identify or quantify specific airborne hydrocarbon compounds released.

1. Refineries do not manufacture products with specific chemical compositions and, therefore, do not routinely measure the chemical composition of their products or emissions. Rather, refinery products have specific properties, such as octane, freeze point, and sulfur content. Crude oil contains thousands of distinct chemicals that are never fully separated during the manufacturing processes.
2. Most hydrocarbons are released through fugitive losses that have small, hard-to-discover, and scattered sources. Because of the large number and wide distribution of potential fugitive emission sources (valves, flanges, pump seals, and tank vents), even a small refinery may have more than 10,000 potentially different sources. Direct measurement of these sources is not practical.

3. The quantities released through any single source are extremely small—on the order of pounds per year—and are dilute and difficult to measure. Total hydrocarbons released from Yorktown Refinery from all sources were approximately 0.3% wt of the total crude oil processed. Therefore, they would not be detected through normal mass balances (National Research Council 1990).

Thus, collecting detailed, chemical-specific release information used to characterize the refinery was expensive and time-consuming. This project developed a sampling and monitoring program that included about 1000 samples. The probable accuracy of most measurements is ± 100 ton/yr. Each sample was analyzed for 15 to 20 chemicals. The sampling program took about 12 months to complete at a cost of about $1 million. Even with this time and dollar commitment, only selected sources were sampled. The final release inventory was assembled using a combination of sampling, measurements, dispersion modeling, and estimates based on emission factors. This effort resulted in the first major database showing all releases from a single facility into all environmental media at one point in time.

Besides analyzing the makeup of these samples, the project team attempted to associate each sample with a source in the refinery. The nature of refinery operations, as well as the complexity of processing equipment, made this task difficult (see figure 3.5.1). Special monitoring methods were used, and some new techniques were developed. For example, ambient air monitoring upwind and downwind of the refinery was used to infer information on fugitive emissions.

RELEASES TO THE ENVIRONMENT

Analysis of the emission data showed that, of the material that was released, most—some 88%—were released into the air. Hydrocarbon vapors made up more than half, with nitrogen, sulfur, and carbon monoxide constituting the second large category. Smaller amounts of material were released to land and a very small amount to water. In fact, one of the findings of the project was the high quality of water around the refinery.

The release inventory process allowed a comparison of pollutant generation, on-site management, and ultimate releases to the environment. The refinery generates about 27,500 ton/yr of pollutants. As

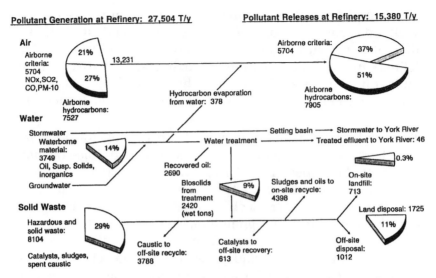

Figure 3.5.2. Pollutant transfers/recycle/treatment within Yorktown refinery.

a result of site hydrogeology, on-site wastewater treatment, and solid waste recycling practices, about 12,000 tons are recovered, treated, or recycled and do not leave the refinery site. Of the remaining 15,500 tons, about 90% are released to the air (note the sum of percentages on the right side of figure 3.5.2). Once this fact was recognized, the workgroup focused much of the remaining project resources on the largest releases—airborne emissions.

Cross-Media Transfers

Modeling studies indicated relatively little naturally occurring transfer of hydrocarbon emissions from air into other media (Cohen et al. 1991). Most hydrocarbons are not very water-soluble and so are not easily removed from the air by rainfall. Although the fate of criteria airborne pollutants (such as NO_x and SO_2) was not studied in this project, they are known to be scavenged by rainfall and can contribute to nitrogen loads and pH changes in lakes and soil. Measurements and modeling results showed small transfers from some surface water ponds to groundwater. Groundwater also enters the wastewater treatment system through the underground sewers, resulting in a net groundwater inflow.

Intentional transfers of pollutants between media do occur, particularly as a result of pollution management activities. Over 370 ton/yr of hydrocarbons initially present in wastewater streams are volatilized into air from the water collection system. More than 2000 ton/yr of biosolids are produced by biotreating wastewater in the refinery's activated sludge system.

Public Perception Survey

One of the unique features of this project developed from the workgroup's desire to include information about York County's perception of the impact the refinery has on the surrounding environment and its role in local environmental concerns. Three distinct activities were undertaken to collect this information: thought leader interviews, two focus group meetings, and a telephone survey of 200 households.

Land development was found to be the major quality of life concern in the Yorktown area. People cited traffic, sewer and water problems, and general development as the major detractors from the current quality of life rather than impacts from the refinery. Another area of concern that surfaced in the focus groups is the general problem of reduced yields of fish and shellfish in Chesapeake Bay, although there is no clear link between the existence of the refinery and the yields. The survey also identified a need for a credible "communication channel" that the refinery and community can use, if needed, since there seems to be no authoritative source of information about environmental problems that people accept as reliable.

PROJECT RANKING

In order to develop a list of emission reduction options, an important step in the Yorktown project design, the information collected during the process of developing a release inventory was examined at a workship attended by about 120 people from Amoco, EPA, and outside organizations. The workshop participants identified some 50 emission reduction options. The workgroup subsequently narrowed this list to 12 options for more careful quantitative analysis. The criteria for selection included those options that were technically feasible now, that offered potentially large release reductions, that addressed different environmental media, and that posed no process or worker safety problems. Projects designed to comply with several current or anticipated regulations were also included.

The workshop also addressed screening criteria to help prioritize the options. Potential barriers to, or incentives for, implementation, along with permitting concerns, were among the criteria considered. The diverse viewpoints brought to all these discussions helped guide subsequent project activities. These views reinforced the workgroup's desire to consider broader issues such as multimedia release management consequences, and future liability impacts. The workshop was able to consider these issues more comprehensively than either government or industry alone would normally do.

The projects were evaluated for such features as risk reduction potential, technical merit, cost, construction safety and operability and then ranked according to weights assigned to these characteristics by individual groups. Table 3.5.1 shows the projects ranked as high, medium, and low, according to these five different criteria. The projects listed in the table are numbered as in figure 3.5.1. It is interesting to note that both Amoco and EPA, using different weights for ranking criteria, chose the same option as the best one in reducing emissions.

The study showed that source reduction options are more cost-effective than treatment but do not necessarily pay for themselves (see table 3.5.2). Note again that the projects are numbered as in figure 3.5.1. The average break-even cost of the source reduction projects was $2.50 per gallon, while the refinery receives an average of $0.75 per gallon for its gasoline product.

Recommended Pollution Prevention Options

The project team identified a set of options that could prevent or capture almost 6900 tons of emission a year for a cost of about $510 per ton. These options included the installation of controls to reduce emissions from barge-loading operations.

At Yorktown, over 80% of the gasoline and other products leave the refinery by barge. Capturing these emissions was found to be the single most *effective* measure although *not the lowest in cost.* Also effective, but less costly, were three other suggestions: (1) the installation of improved seals on selected storage tanks; (2) instituting a leak detection and repair program to reduce small leaks around valves and flanges; and (3) the upgrading of emergency venting equipment, called blowdown stacks, to reduce hydrocarbon losses to the air.

The annual average concentration of benzene in the vicinity of the Yorktown Refinery were plotted based on a computer-generated

Table 3.5.1. Yorktown Project pollution prevention option scores by ranking technique

No.	Project	Release reduction	Benzene exposure reduction	Annual cost	Cost-effective release reduction	Cost-effective exposure reducton	Analytical hierarchy process
1	Reroute desalter effluent[a]				M	M	
3b	Install electrostatic precipitator at FCU[b]	M			M	M	
4	Eliminate coker pond[c]				M	M	
5c	Secondary seals on gasoline tanks with floating roofs[c]	M	M		H	H	M
6	Soils control	M		M	H		
7A	Blowdown system upgrade[c]	H	M		H	M	
7B	Drainage system upgrade						
7C	Wastewater treatment plant upgrade						
8	Modify sampling procedures			H	H		
9	Barge loading vapor control[c]	M	H		M	H	H
10	Sour water system improvements			M	M		
11b	Quarterly LDAR[d] program at 10,000 pp	M		M	H	H	

[a]Blank entries indicate low (L) ranking.
[b]Fluidized cracking unit, a catalytic process unit that produces gasoline components.
[c]Options with high (H) and medium (M) rankings for all but one criterion.
[d]Leak detection and repair, a systematic program to identify and repair minute leaks on piping components.
[e]Numbers in the first column refer to stages in the refinery process as identified in figure 3.5.1.

Table 3.5.2. Yorktown Pollution prevention project—cost effective release reduction ranking

No.	Project	Pollutant	Pollution prevention mode	Cost-effectiveness $/ton[a] of pollutant reduced	Recovery cost[b] ($/gal)
11b	Quarterly LDAR[c] program at 10,000 ppm	VOCs[d]	Source reduction	270	0.88
5c	Secondary seals on tanks with floating roofs	VOCs	Source reduction	287	0.93
7A	Blowdown system upgrade	VOCs	Treatment	320	1.04
6	Soils controls	Hazardous waste	Source reduction	383	
8	Modify sampling procedures	VOCs	Source reduction	429	1.39
9	Barge loading vapor controls	VOCs	Recovery/treatment	2,094	7.00
4	Eliminate coker cooling water pond	VOCs	Source reduction	4,862	16.00
1	Reroute desalter water	VOCs	Recycle	6,279	21.00
3b	Install electrostatic precipitator at FCU[e]	Catalyst fines	Disposal	8,106	
10	Sour water system improvements	H_2S,[f] ammonia	Recycle/treatment	11,056	
7B	Drainage system upgrade	VOCs	Treatment	52,809	171.00
7C	Wastewater treatment plant upgrade	VOCs	Treatment	127,638	415.00

[a]Annual cost of reducing one ton of the associated pollutant, including operating, maintenance, and annualized capital costs.
[b]Price of gasoline required for the dollar value of the emissions reduced to equal the cost of the emission reduction project.
[c]Leak detection and repair, a systematic program to identify and repair minute leaks on piping components.
[d]Volatile organic compounds, that is, volatile hydrocarbons.
[e]Fluidized cracking unit, a catalytic process unit that produces gasoline components.
[f]Hydrogen sulfide, a colorless, toxic gas.
[g]Numbers in the first column refers to stages in the refinery process as identified in figure 3.5.1.

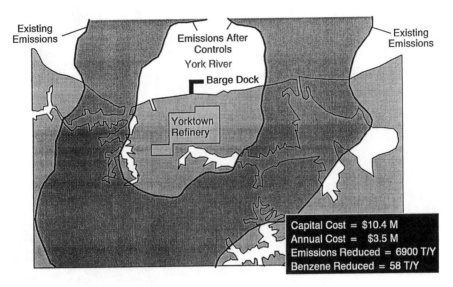

Figure 3.5.3. Recommended controls at Yorktown refinery.

model of how local weather patterns distribute emissions at this location (see figure 3.5.3). By adopting the recommended options developed from the study, emissions could be reduced by the amount shown between the line marked *Existing Emissions* and the line labeled *Emissions After Controls.* The cost of this significant improvement would be a capital investment of about $10 million, or about $510 per ton.

However, the Yorktown Refinery faces some mandated require-ments, such as the modification of its sewer system. The cost of these required modifications is four times greater than the recommended options, or an estimated capital cost of $41 million, equivalent to about $78,000 per ton. Figure 3.5.4 shows the minor emissions improvement that will be gained as a result of upgrading the sewer system. The narrow band between the line marked *Existing Emissions* and the line labeled *Emissions After Controls* demon-strates the shortcomings of imposing industrywide measures without taking into account how effective they will be at a particular facility.

In the case of Yorktown, mandated requirements for all sources, including the sewer system, will reduce emissions at a cost of $2,400 per ton, while optional alternatives could achieve virtually the same emission reductions at a cost of about $500 a ton.

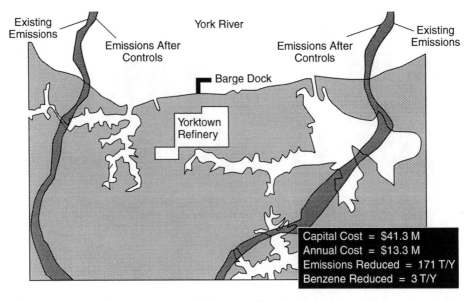

Figure 3.5.4. Mandatory sewer controls at Yorktown refinery.

OVERALL RECOMMENDATIONS

In addition to fairly specific pollution control options, the Yorktown Project identified several broad policy recommendations aimed at achieving an improved environment.

Encourage Long-Term Compliance Solutions

Because mandatory compliance deadlines are short, the current system is directed at short-term fixes, sacrificing more effective, if less immediate, solutions. Most programs require compliance within six months to three years. However, the design, engineering, and construction of many environmental improvement projects may take much longer than the compliance deadlines allow. In addition, there are often delays associated with difficulties in interpreting regulations, understanding design criteria, obtaining construction permits, and developing unfamiliar technologies. In light of these conditions the Yorktown Project team recommended that legislators and regulators adopt more realistic time frames to encourage long-term solutions.

Encourage Innovative Approaches

Since the initiation of environmental legislation in the early 1970s, regulations have maintained a narrow focus on single issues. This piecemeal approach lacks an overall goal that might be achieved through a variety of programs. Better coordination among numerous environmental requirements could help industry develop broader management initiatives to meet the overall objectives. In response to these findings, another recommendation calls for the introduction of incentives to conduct facilitywide assessments and emission reduction strategies.

Develop Better Data on Emissions

The Yorktown Study revealed that some regulations are misdirected, imposing controls on particular refinery releases that are no more than minor sources of pollution. Ineffective regulations arise from a lack of sound, reliable data that accurately identify the type, amount, and source of emissions. Such a database is a necessity for cost-effective pollution control. The Yorktown Project team recommended that additonal research be undertaken to develop improved techniques for data collection, analysis, and management.

A need for better data, however, does not constitute a need for simply more data. The project pointed out the numerical limitations of the Toxics Release Inventory (TRI) as a tool to measure environmental improvement. The TRI is further limited by its focus on pounds of emission reduction, without consideration of the corresponding risk reduction. The Yorktown study focused on risk reduction to evaluate a pollution prevention option's effectiveness. The use of such a measure can be translated into a quantitative goal if one develops an appropriate risk reduction "stopping point," a determination well beyond the scope of this project.

Encourage More Government-Industry Cooperation

The Yorktown Project is the first joint effort between government and industry to study pollution control opportunities at an operating refinery. Many benefits flowed from the project, including greater appreciation on both sides for the problems of the other. With a spirit of open-mindedness and cooperation, a better knowledge of environmental problems, along with more innovative, cost-effective solutions, can be developed. A fourth recommendation of the project is to encourage additional partnerships between the public and private sectors.

YORKTOWN FOLLOW-UP ACTIVITIES

The Yorktown study has ended. The project produced a great deal of knowledge for both industry and government agencies. What have we accomplished as a result of everyone's efforts? First, let's review some specific Yorktown follow-up activities, including the implementation of projects identified in the project, some subsequent emission data gathering, and an analysis of environmental laws.

Implementation of Pollution Prevention Projects

The Yorktown Project was designed to inventory refinery emissions, develop options to reduce those emissions, and analyze the regulatory system for incentives for, and barriers to, implementation. The study identified some new and effective pollution prevention ideas. At the conclusion of the study, however, no mechanism existed to utilize cost-effectiveness to determine which pollution prevention projects should be implemented. The facility was required to comply with all the regulations in effect.

Several of the projects identified in the study have been engineered and installed, some at little cost and others at high cost. The upgrades to the sewer system and wastewater treatment plant to reduce benzene emissions have been completed. This project involved the construction of a new sewer for process wastewater, built completely above ground, and the replacement of the oil/water separator with an aboveground, closed unit. The project was installed at a capital cost of $29 million compared to the estimated $41 million. Secondary seals are being added to the gasoline and crude oil tanks on a multiyear schedule.

Some projects were easier to implement than expected. As a result of some innovative refinery process engineering, the use of the coker cooling pond was completely eliminated (rather than only controlling emissions from it), doubling the emission reductions at a fraction of the cost.

On the other hand, the project that offered the most risk reduction potential—controlling emissions during barge loading—is still on hold. The capital funds commitment remains in the current investment plan, but the engineering awaits the issuance of the regulations that address this emission source. The refinery cannot afford to chance implementing a system to control the emissions that may not comply with a future technology or performance standard.

Subsequent Emission Data Gathering

One of the project findings was the need for good emissions data for development of regulations and facility-specific pollution prevention plans. The results of subsequent data-gathering efforts paint a much different picture about the refinery emissions than previously envisioned (see figure 3.5.5). First, since the testing of the blowdown stack emissions proved very difficult during the project, Amoco continued to investigate other available sampling and analytic techniques to refine the blowdown stack emission estimates. As a result of that investigation, a blowdown stack sampling method that uses a special tracer gas was developed and employed. Based on the subsequent stack testing, volatile organic emissions from this source are now estimated to be 2800 ton/yr, a reduction of 5100 ton/yr over the previous analysis. Among these, benzene emissions from the facility are currently estimated at 35 ton/yr, 32 ton/yr lower than the estimate in the original study.

Second, a partial survey of the refinery has been completed to identify and inventory the thousands of small valves and connectors that are the source of "fugitive" air emissions. With the aid of this

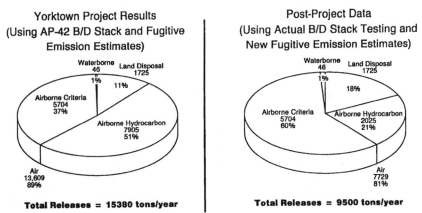

Figure 3.5.5. Total releases entering the environment from the Yorktown refinery (in ton/yr).

screening of refinery fugitive emissions source components and the corresponding estimation technique, fugitive emissions are now estimated at one-fifth of the originally reported emissions (Amoco/EPA 1994). The Radian Corporation has recently completed a comprehensive industry survey of refinery fugitive emissions for the Western States Petroleum Association and the American Petroleum Institute (WSPA/API 1993). This study indicated that the current EPA correlation factors overestimate fugitive emissions by a factor of 10 to 15. Therefore, the Yorktown Refinery fugitive emissions, a subcategory of total emissions, are now estimated at 16 ton/yr, approximately one-fiftieth of the originally reported 796 ton/yr, or a reduction of 780 tons. According to these new data, a formal leak detection and repair program would not be cost-effective. Such a program may be required for compliance with upcoming refinery maximum achievable control technology (MACT) guidelines. Thus, based entirely on better emissions data from the two categories of sources, aggregate airborne hydrocarbon emissions are now estimated at 2025 ton/yr.

Analysis of Major Federal Environmental Laws

The study showed that, given flexibility to incorporate site-specific factors, a facility could develop an alternative compliance strategy that could achieve equal or greater environmental benefits at lower cost. But where is the flexibility needed? As a follow-up to the project, Amoco and outside counsel completed a review of major federal environmental statutes, identifying over 20 procedures for implementing more flexible and innovative approaches (Raffle and Mitchell 1993). Many of these procedures are available within existing laws, but the EPA must take the lead in encouraging their maximum use.

OPPORTUNITIES FOR MARKET-BASED APPROACHES

The current system of environmental regulations can be credited with driving significant improvements in the quality of our air, water, and land over the past two decades. But, as the Yorktown Project demonstrated, the current system provides few incentives to achieve much more than mere compliance. Fortunately, a consensus seems to be emerging that an alternative approach to the current command-and-control system might achieve environmental benefits more effectively. In fact, the SO_2 trading program, which was created as part of the Clean Air Act Amendments of 1990, seems to be working well, and shows real promise of achieving the performance goals set by the law while allowing companies flexibility to coordinate environmental improvement plans with business plans.

Other opportunities, potentially more effective than current regulations, exist for market-based approaches to be used to achieve environmental improvements. Let me mention three.

Early Emission Reduction Banking Systems

Of the pollution prevention options identified in the Yorktown Project, the one that offered the most potential to reduce the exposure from benzene emissions was the controls on marine vapor loading. Yet, because the area has good air quality, rules for these controls are years away from development. Currently, facilities have an incentive to delay potential emission reductions so as, effectively, to bank them internally for offsets that may be required for future construction of new facilities or increases in production rates.

Establishing an authorized banking system for early emission reduction would be a real incentive for facilities to implement projects voluntarily before regulations require them. In fact, a banking system may be an incentive for companies to develop comprehensive, multimedia emissions inventories at their facilities, similar to those of the Yorktown Project, in order to identify the most cost-effective ways to reduce specific emissions.

Stationary Versus Mobile Source Emissions Trading

In some cases, stationary sources and mobile sources emit the same kinds of pollutants in the same geographic areas. Opportunities to find the most effective ways to reduce these emissions are not always obvious, nor are these opportunities always permitted.

For many facilities, equivalent emission reductions can be achieved much more readily at stationary sources than by developing complex employee commute options. Since the objective of the stationary and mobile source programs is the same, even though the sources are somewhat different, emission trading between mobile and stationary sources should be evaluated to identify the potential to achieve emission reductions in the most expeditious, effective manner.

Preference to Pollution Prevention Solutions
Versus End-of-Pipe Controls

Environmental regulations often stipulate the types of technology to be applied to control emissions from specific sources. As a result, the power of the technology market is not tapped. Regulations sometimes specify performance standards, but these are often based on implementation of a certain control technology. While the emphasis on performance standard is typically preferred by industry, the implication of the underlying technology can often limit the use of alternative approaches. Incentives can help stimulate development of new pollution prevention technology.

In some water discharge case studies, certain compliance strategies that achieve emission reductions through source reduction were found to be incapable of achieving the level of emission reduction achieved by end-of-pipe treatment and required for compliance with the water quality standard. However, for source reduction options, the energy use may be lower, additional waste generated by treatment is avoided, and waste during construction and maintenance activities is avoided. When evaluated on a multimedia basis, the overall benefits of the source reduction option may outweigh the end-of-pipe treatment. Yet, if the decision is based solely on single medium compliance, the advantages of source reduction could be lost.

For control of volatile organic air emissions, the performance standard is often set at 95%–98% efficiency. Yet the universal control technique that achieves 98% efficiency is combustion or incineration. Incineration recovers none of the product or stream emitted and produces nitrous oxides, which may have a greater impact on the formation of ground-level ozone.

In these cases, source reduction benefits could be achieved if an incentive could be provided to make source reduction more valuable than end-of-pipe controls. Such a credit system could stimulate additional research for source reduction technologies to be developed and implemented.

CONCLUSION

The Yorktown experience demonstrates the opportunities and pitfalls that can occur when government and industry work together. The opportunities are significant. The pitfalls are worth overcoming.

All organizations—U.S. EPA, the Commonwealth of Virginia, and Amoco—sought to develop and test innovative approaches to reduce releases to the environment. In general, we found that opportunities exist at this facility. However, the findings of one study cannot become the basis for setting national policy.

Given the large potential for achieving better environmental benefits more cost-effectively, we believe that additional demonstration projects should be undertaken to develop more innovative and more cost-effective approaches to environmental protection. In the debate for the reauthorization of the Clean Water Act during 1994, an amendment was added to allow as many as 10 demonstration projects to be undertaken to craft multimedia release management strategies, thereby providing industrial facilities with an incentive to conduct comprehensive facilitywide emission assessments such as the Yorktown Project. If it had been enacted into law, this mechanism could have generated information on how to achieve better results at lower costs.

The EPA is moving in the right direction with two of its own initiatives, the Common Sense Initiative and the Sustainable Industry Project, which are attempting to find collaborative approaches to achieve "cleaner, cheaper, faster" solutions to environmental problems. We hope that these initiatives will evaluate and expand market-based systems to create incentives for innovative technology choices. Most importantly, these systems will help unleash the creativity of individuals in industry and government and among the public to jointly address the common goal of protecting the environment.

REFERENCES

Amoco/EPA. 1994. *Supplemental Report.* Prepared by Ronald E. Schmitt for the Amoco/U.S. EPA Yorktown Pollution Prevention Project. Chicago, Il: Amoco Corporation.

API. 1991. *The Generation and Management of Wastes and Secondary Materials in the Petroleum Refining Industry, 1987–88.* Washington, DC: American Petroleum Institute.

Cohen, Y., Allen, D. T., Blewitt, D. N., and Klee, H. 1991. *A Multimedia Emissions Assessment of the Amoco Yorktown Refinery.* Presented at the 84th Annual Air and Waste Management Association meeting, Vancouver, BC, Canada. June 17.

National Research Council. 1990. *Tracking Toxic Substances at Industrial Facilities: Engineering Mass Balance Versus Materials Accounting.* Washington, DC: National Research Council, National Academy Press.

Raffle, B. I., and Mitchell, D. F. 1993. *Effective Environmental Strategies: Opportunities for Innovation and Flexibility Under Federal Environmental Laws.* Chicago, IL.: Amoco Corporation.

WSPA/API. 1993. *1993 Refinery Fugitive Emission Study.* Prepared by Radian Corporation for the Western States Petroleum and the American Petroleum Institute. Sacramento, CA: Radian Corporation.

3.51

❏

DISCUSSANT

Frances H. Irwin

The directors of the Yorktown Project deserve much credit for their patience and their persistence. It was not an easy project. It reflects well on both Amoco and EPA that further demonstration projects are being undertaken.

I will address three points in discussing the project and market incentives from the perspective of the World Wildlife Fund (WWF): first, the nature of the conservation challenge; second, some lessons learned; and, third, directions for future demonstration projects.

THE NATURE OF THE CONSERVATION CHALLENGE

Environmental problems take three forms. First are the relatively well understood problems, such as lead poisoning and ozone depletion. Second are the problems we know exist but don't know how to assess and solve. The effects of interactions among pollutants are an example. Then, third are the problems we don't know about. From experience, however, we are fairly certain that they will emerge.

It is difficult to use market-based approaches like emissions trading for dealing with any but the first type—the problems that we understand relatively well. For the others, we lack a framework and required data. Environmentalists tend to worry about the second and third kinds of problems. The WWF's Wildlife and Contaminants Program, for example, is deeply involved in developing a new understanding of how some chemicals, sometimes at very low levels, can act like hormones and change the way an embryo develops.

311

Wildlife populations of fish, birds, and mammals are affected. In humans, exposure to chemicals that act like hormones is hypothesized to be linked to a whole range of effects from immune dysfunction and infertility to several kinds of cancer. We do not yet test for these effects because we are only beginning to understand them.

I have gone into detail on an example of a problem that is perhaps halfway between the second and third types described above to explain one reason why economic instruments, important as they are, can play only a limited role in environmental policy.

LEARNING FROM THE AMOCO PROJECT

What did we learn from the Amoco Project? I would emphasize four learnings.

First, involving more of the players does result in better solutions. It is worth the struggle to do so. All kinds of boundaries had to be crossed, both within Amoco and within EPA and between the company and the agency. Some were more successfully crossed than others. Most of the energy went into making those links rather than in working with environmental and community groups. I appreciated the opportunity to participate in the peer review group. However, despite much discussion inside and outside that group, neither advocacy groups nor local groups were invited to participate directly in the Yorktown Project. A real success of the project is Amoco's new willingness to work with more activist-type groups at both the local and national level as it engages in another round of demonstration projects.

Second, existing information requirements don't provide the data needed to make decisions. We learned that a facility and EPA both need much better information on the pollutants the facility releases. This information is crucial to defining how to get the most risk reduction at the least cost.

Third, we need both to assess and to develop prevention and control measures for "whole" facilities and the "whole" environment. The governmental structure for environmental management has been fragmented among a whole range of separate programs with, at least, these six consequences:

1. Pollution control methods that move pollutants from one part of the environment to another are encouraged; for example, incin-

erating waste results in air pollution and ash that must be landfilled.

2. Problems are not accurately identified and thus not effectively managed. At the Amoco refinery, capturing emissions at the barge-loading facility turned out to be the single most effective measure. This had not been identified before because no environmental program had ever asked about those emissions.
3. What is often the best solution both economically and environmentally—source reduction—is ignored. The Amoco project identified some source reduction options, much to the surprise of some participants.
4. Priorities among environmental problems cannot be set—or implemented. Current laws may make it difficult to act on the most important problems first.
5. Government environmental management by program makes it difficult to integrate environmental policy into other policy sectors, such as industry, agriculture, and transportation.
6. Fragmented programs result in an excessively complex and inconsistent administrative structure (Irwin 1992).

Fourth, prevention opportunities do exist and often pay for themselves. This was also a hard-fought issue in the project, and I am pleased to see how Amoco now presents it. During much of the project, there was resistance to the idea that one could find *any* source reduction options at this refinery. That they were found demonstrates that the hierarchy of environmental management in the Pollution Prevention Act, which puts source reduction first, is a helpful policy in changing expectations.

My own view is that future demonstration projects should clearly address both source reduction and control. As this project concludes, solutions include both types of measures. Neither should be ignored, but the hierarchy should be used.

A NEW FRAMEWORK TO INCREASE COMPANY CHOICE AND CONSERVATION RESULTS

Can we increase the company's choice in the means and timing of achieving conservation and environmental goals by increasing public accountability? One way to do this is to set environmental goals and track achievement through an expanded version of the Toxics Release Inventory. Economic incentives can then be used as one tool to achieve the goals.

The Amoco project begins to demonstrate how this can be done at a facility. The U.S. 33-50 project, under which companies reduced their releases of 17 chemicals by 50%, has also shown the possibilities as well as some of the pitfalls of such an approach. Another example at the national level is the Dutch National Environmental Policy Plan. It sets goals for reduction as high as 90% for some pollutants. To set and track goals, we need a measurement system. The multimedia facility assessment that Ron Schmitt mentioned would be one way of moving toward such a system at the facility itself. It needs to be linked to national reporting requirements that provide a common format so that data can be compared across facilities and countries (Irwin et al. 1995).

Demonstration projects are a good way to move toward a new pollution system. Manik Roy will talk about the Great Printers Project. The WWF has an agricultural pollution prevention project in the Great Lakes that is investigating how pesticide reduction goals can be set and progress measured. At the national level, WWF is taking the lead in drafting legislation to establish a cohesive national program to motivate users to reduce their use of, and reliance on, pesticides. Half of water pollutants in the United States come from non-point sources. Thus, it is very important that we don't look just at the industrial sector.

Future demonstration projects should develop and evaluate policy tools such as pollution prevention planning to increase efficiency in the use of energy and materials and to reduce the use of toxic chemicals. Future projects might well put more emphasis on linking what happens at the facility to the sector and regional level and include longer time frames both for changes in technology and effects on the environment.

There is a real possibility, in the next four or five years, that we may reinvent how we prevent and control pollution. If we reframe fragmented air, water, waste, and toxics programs into a coherent approach, set clear environmental goals, and require public reporting, market incentives are one important way to achieve those goals.

REFERENCES

Irwin, Frances H. 1992. "An Integrated Framework for Preventing Pollution and Protecting the Environment," *Environmental Law* 22: pp. 1–76.

Irwin, Frances H., Tom Natan, Warren R. Muir, Eric S. Howard, Lois Lobo, and Sharon Martin. 1995. *A Proposed Benchmark for Reporting on Chemicals at Industrial Facilities.* Washington DC: World Wildlife Fund.

3.52

❑

DISCUSSANT

*Mahesh Podar**

❑ The joint pollution prevention project that EPA and Amoco began in 1990 marks the first time that a regulated company volunteered, and a regulating entity agreed, to try a new approach. Details about various aspects of the project can be found in the 24 reports, available through the National Technical Information Service, that are summarized in *The Project Summary* (Podar and Klee 1992).

As EPA's project manager, I was actively involved in the day-to-day activities of the project. I would like, first, to discuss some unique features of this project, then to describe activities stimulated by Yorktown, and, finally, to comment on market incentive activities that EPA is pursuing.

At the outset, both EPA and Amoco agreed that the adversarial approach to carrying out the mission of protecting the environment was highly inefficient and that we had to try something different. We found that, by taking local conditions into account, we could find solutions that were tailored to the problems yet guided by some national baseline of minimum requirements. We found that, if the refinery was given a goal, for example, reducing the releases to the environment by 90%, it could meet that goal, which could be long-term, more cost-effectively than by achieving reductions to meet specific regulatory requirements. At the outset, we realized that, without information, it would be difficult to reach any informed conclusions, and we soon realized that we did not have as much

*The views expressed in this paper are those of the author and should not be construed as the official policy of the U.S. Environmental Protection Agency.

315

information as we thought we had. We also had collected some information we did not need. The generation of additional relevant information became a vital part of the project.

Since the completion of the joint pollution prevention project, EPA has begun a number of different activities that attempt to further the findings of this study. The Common Sense Initiative (CSI) builds on Yorktown. The CSI has begun with six industries — auto assembly, computers and electronics, iron and steel, metal plating and finishing, oil refining, and printing — in which a partnership approach will find opportunities for environmental improvement in six key areas:

1. Regulation. Review existing regulations and those that are in the pipeline to identify opportunities to achieve better environmental results at less cost and to improve new regulations through more coordination between various media requirements.
2. Pollution prevention. Make pollution prevention the guiding principle and a basis for doing business. Pollution not generated does not have to be treated and disposed.
3. Information collection. Make information collection easier for the industry and more accessible to the public. This would include use of electronic databases and integration of various media-specific information to eliminate duplication.
4. Strong enforcement. Increased flexibility in achieving environmental results will be accompanied by strong enforcement to ensure that there are no competitive disadvantages associated with compliance.
5. Improved permitting process. Provide greater opportunity for public participation and a faster, more timely response to industry.
6. Encouragement of innovative technology. Promote creativity and innovation by ensuring that EPA does not specify technology in its standards.

Different CSI industries are at different stages, but some issues common to all six industries are beginning to emerge. For example, concerns about how to bring land at abandoned plants, called "brownfields," back into productive uses are being actively discussed. All six industries are addressing the issue of increasing the efficiency of data collection as well as increasing public access. All six industries have identified the need for improving the permitting process to expedite the issuance of needed permits as well as increasing public participation in the process.

In addition to changing the way it does business, EPA is also promoting economic incentives. Examples of environmental markets to control acid rain, urban ozone, and greenhouse gases are discussed in this book. Regarding water, this incentive takes the form of effluent trading. Five approaches to effluent trading are being explored and promoted: effluent reduction trading between point sources, discharge pipes within a facility (intraplant), between point and non-point sources, and between non-point sources, and pretreatment trading, that is effluent reduction trading between discharges to the same publicly owned treatment work (POTW). These approaches are described in a report by Podar and Luttner (1993). In February 1996, EPA published a policy statement promoting trading. In May 1996, EPA published a *draft framework for watershed-based trading.* This is a how-to manual, not a cookbook, that will be revised to incorporate public comments.

In summary, flexible, innovative approaches provide us with an opportunity to build on the achievements of the last 25 years. These approaches will enable us to develop cost-effective solutions to local problems by building on the base of minimum national standards. Market incentives will further enhance the efficiency of these solutions. Flexibility and incentives will also encourage development of innovative technology and take advantage of American ingenuity to reach the goal of a safe environment for us to live in.

REFERENCES

Podar, Mahesh, and Howard Klee (1992). *Amoco/USEPA Pollution Prevention Project: Project Summary.* NTIS PB2228527. Washington, DC: National Technical Information Service. June.

Podar, Mahesh, and Mark Luttner (1993). "Economic Incentives in the Clean Water Act: Some Preliminary Results." Presented at the 84th Annual Meeting of Air and Waste Management Association, Denver, CO, June 13–18, 1993.

3.53

❏

DISCUSSANT

Manik Roy

❏ I worked at EPA at the time of the Yorktown experiment and was a member of the EPA advisory committee to the project. Yorktown has spawned lots of offspring: the President's Commission on Sustainable Development has been looking into setting up demonstration projects following the Yorktown model; the Senate Environmental and Public Works Committee developed legislation based on the same approach; and the EPA, in its own vein, is running out ahead of both of those groups to develop the idea. So, I think Yorktown has led to discussion of a lot of extremely important issues.

I'm not sure, however, that what I got out of the project was a big argument for market-based incentives. Yes, there is a need for flexibility in the way we encourage companies to meet environmental goals, but there are ways other than environmental markets to provide incentives. I think the incentive that worked in Amoco's case was a large company's interest in being seen as a good corporate citizen.

Another reason that I urge caution is that market trading of pollution permits raises social value questions not easily quantifiable in terms of benefit-cost analysis, the traditional justification for trading. For example, suppose two different technologies with the same expected cost reduce pollution to the same extent. The first improvement has a better expected value to human health and a better expected value to the current generation, but the outcome is highly uncertain. The second has a better expected value to the ecosystem, a better expected value to future generations, and the outcome is less uncertain. How can we decide which improvement to choose? Should we favor the present over future generations? Should we engage in risk-averse or risk-neutral solutions in the face of uncertainty in controlling pollution? These are not scientifically determinant issues; they are issues of social values. And answers to these questions, in our political system, can be achieved by only posing the problems to the general public and involving them in the decision process.

As we move into demonstration projects, the children of York-town, I would like to see a more forthright wrestling with these social values by getting the communities involved. Get the community that lives around the facility involved, get the environmental groups who have been opinion leaders on these sorts of issues involved, get the people who work there involved—get all the stakeholders involved in a meaningful discussion of the social values vital to environment-al policy. Pollution control is not just a question of coming up with the right numbers.

THE GREAT PRINTERS PROJECT

One of the most important examples of using market tools to promote pollution prevention is the Great Printers Project. The market tools used here overcome the shortcomings of information in the markets faced by the typical printer.

The Great Printers Project, started in 1992, is a partnership of the Council of Great Lakes Governors, Printing Industries of America, and the Environmental Defense Fund (EDF), with strong support from the U.S. EPA (EDF 1994). In July of 1994, these partners, along with the coalition of printers, environmentalists, regulators, assist-ance providers, and labor representatives that make up the Great Printers Project team, released a series of recommendations for establishing pollution prevention as the standard business practice in the lithographic printing industry. The recommendations chal-lenged every participant in the printing process—printers, print buyers, suppliers, distributors, government regulators, and technical assistance organizations—to consider proactively the impacts of printing operations and to work with printers to protect the environ-ment.

Since that time, implementation activities have been under way in four pilot states with demonstrated records as innovators: Illinois, Michigan, Minnesota, and Wisconsin. The implementation effort has been a collaboration of groups initially involved in crafting the Great Printers Project recommendations, as well as state, environmental, industry, and technical assistance partners in each of the four pilot states. The U.S. Environmental Protection Agency has strongly endorsed and supported the Great Printers Project since its inception in 1992.

The Great Printers Project focuses on supporting print shops committed to furthering *Great Printers principals* through special enrollment programs. Key components of these programs include:

- Informing print buyers (through printers themselves) and thereby generating customer demand to create a market for environmentally superior printing.
- Improving access to information on technology and financial resources for printers.
- Simplifying governmental requirements so that the printer can readily understand, meet, and exceed his or her environmental obligations, primarily through pollution prevention.

Generating Customer Demand

To create a demand for Great Printing, the Great Printers Project recommends that print buyers work in partnership with printers to specify their requirements in ways that produce quality print jobs that do not compromise the environment.

Two of the four pilot states, Michigan and Wisconsin, have recently launched campaigns to enroll printers committed to the Great Printing principals and to encourage customers to "Buy Great Printing." Similar efforts are under way in Illinois and Minnesota. By the summer of 1997, the four states expect to enroll 500 Great Printers.

Improving Access to Information on Technology and Financial Resources

In 1994, the Great Printers Project recommended that the EPA establish a national resource center to provide reliable, up-to-date information specifically on the printing industry. The center would be used by organizations and individuals that provide business assistance to printers and would:

- Respond to calls from business assistance providers for information.
- Provide information on printers' regulatory requirements.
- Have a process to keep current on new technologies and disseminate relevant information on costs, savings, and environmental benefits of new technologies.
- Identify research and development needs for the industry.
- Foster information exchange among service providers and enable peer-to-peer networking.
- Disseminate relevant published material.
- Train business assistance providers on pollution prevention opportunities in printing.
- Continually evaluate its own usefulness.

With over $500,000 in funding from the EPA, the Printers National Environmental Assistance Center (PNEAC) was launched in 1995, fulfilling a commitment by EPA administrator Carol Browner and President Clinton to establish small-business compliance assistance centers. In addition to the EPA, PNEAC partners include the Illinois Hazardous Waste Research and Information Center (HWRIC), the University of Wisconsin Solid and Hazardous Waste Education Center (SHWEC), the Environmental Defense Fund (EDF), Printing Industries of America (PIA), the Council of Great Lakes Governors, and the Graphic Arts Technical Foundation (GATF).

This state-of-the-art "virtual" center is a unique partnership electronically linking pollution prevention technical assistance organizations and printing trade organizations to efficiently provide the most current and complete compliance assistance and pollution prevention information to the industry. The center is designed to complement—not replace—existing service providers. In addition, PNEAC's users include printers and print industry vendors. With two listservs, Printech and Printreg, PNEAC allows industry experts, technical assistance providers, and others to exchange the latest technical and regulatory information with one another.

Cutting the Red Tape

In 1994, the Great Printers Project recommended that government regulators develop simplified regulatory requirements so that the printer can readily understand, meet, and exceed his or her environmental obligations. Since then, the pilot states have been making strides to simplify regulatory requirements applicable to the printing industry.

For example, the state of Wisconsin has been working closely with the EPA to identify and address barriers to consolidated reporting. The Wisconsin Department of Natural Resources has consolidated its multiple air, hazardous waste, and Toxic Release Inventory reporting forms into one simple electronic format that was sent to all Wisconsin businesses in December 1996 — capturing over 90% of the information printers are required to submit to the state annually. In addition, Wisconsin printers will receive an enhanced version of the system that provides information on technical assistance and pollution prevention opportunities, thus taking advantage of the "teachable moment" when printers are most aware of their regulatory obligations and are open to receiving information on ways to minimize the environmental impacts of their shop operations.

By 1997, all four pilot states intend to develop and implement an improved reporting system to allow printers to satisfy their environmental obligations.

Summary: Policy Tools That Create Better Information

The public policy tools being developed by the Great Printers Project provide print buyers better information on the environmental consequences of their product specifications and give printers easier access to information on their environmental obligations and on their options for fulfilling and exceeding those obligations. Environmental information has a cost that is often prohibitive for printers. The Great Printers Project, by reducing that cost, is making it easier for printers to make decisions that are better for business and the environment.

REFERENCES

Environmental Defense Fund in collaboration with The Council of Great Lakes Governors and the Printing Industries of America. 1994. *The Great Printers Project*. Washington, DC. Environmental Defense Fund. July.

Part

4

❑

FINAL MATTERS

❑

4.1

□

CONCLUSIONS:
PERFORMANCE AND PROSPECTS
FOR MARKET-BASED APPROACHES

The Editors

□ The calculations we draw from the studies in this volume are made against a background of increasing remonstration about the character and cost of traditional regulation. For the economy in general, cleanup costs, mainly to be incurred by the use of traditional regulation, are projected by the U.S. EPA to increase to well over 2% of gross domestic product (GDP) by the year 2000; and that estimate is conservative, based as it is on private compliance costs alone (U.S. EPA 1990). Our interpretation of this estimate is that we are moving up the more steeply rising portion of the aggregate marginal cleanup cost curve by using mostly command-and-control (CAC) measures, and we have not yet achieved our environmental goals. Furthermore, new problems loom on the horizon. Not only concerns about the costs of centralized regulation but also complaints about its intrusive nature are increasing among more and more of the interested communities. This discomfort with CAC regulation can only heighten the debate between those who would lower our environmental goals and those who would strive to meet them by cheaper and better ways to secure the cleaner environment that we want.

What can we conclude about the performance of market-based approaches to environmental policy to date and the prospects for their future that bear on this debate? If we are in a period of transition from one regulatory regime for environmental improvement to another, with a new ultimate mix of control measures as the destination, what can be our conclusion about the place of these innovative incentive systems in the future mix? We have had much to say in this book about the strengths and limitations of market-based approaches and especially environmental markets. Can market incentives make a significant contribution by putting us on a different and lower cost curve of control and by giving us a more decentralized and less intrusive relationship between the regulated and regulating communities? Can they do this without, at the same time, sacrificing our environmental goals? The editors have already revealed their hand as supporters of incentive systems and intend to make the case for them. However, they cannot, as the discerning reader cannot, ignore the problems in the design and implementation of incentive systems that have been raised by the Workshop on Market-Based Approaches to Environmental Policy and discussed by our contributors.

Again and again in this book, mention has been made that new market tools, in order to work well in a democratic framework, require an understanding and a complex series of explicit and implicit agreements among the regulated and regulating communities and among other concerned communities. Few of our contributors would claim that these agreements are securely in place. The Workshop on which this book is based made its effort to contribute to that understanding and agreement. The editors detected a movement, during the years of the meetings, toward greater understanding and support of incentive systems among those business, government, environmental, and academic participants who initially may have been resistant to them. With more evidence of performance at hand, with more knowledge of problems and their possible resolution, and with more open exchange of views, the participants, in general, appeared to gain confidence in the future of these innovative control measures. Such movement augurs well for increased future support.

Turning to specific problems revealed in applications, the flag-ship sulfur dioxide cap-and-trade market, several years in operation, provides highly relevant examples of questions not yet fully answered. The quantities of allowances traded at first glance do not seem sufficient to realize the full measure of potential savings. The prices at which they trade suggest that some puzzles have not yet been resolved about utility behavior in this market. Are high trans-action costs a barrier? Is the compatibility of environmental markets with traditional regulation of all sorts raising complications? Witness the difficulties of obtaining clarifying regulatory decisions on market transactions from state and federal regulatory agencies, a lack of clarity that may be impeding utility trading. Has the discriminant type of auction called for by the U.S. Congress constrained and biased private offers and bids?

These difficulties, which have been discussed by our contributors, give us pause and call for further inquiry, but they should not divert us from the progress achieved in this young market, as also recorded by our authors. Hundreds of transactions involving millions of dollars have occurred, resulting in gains from trade that add up to appreciable savings. Our leadoff study put control cost savings at $700 million per year at present, with more to be realized when the comprehensive Phase II begins in the year 2000.

The prices for a ton of emissions revealed in the annual auctions have indeed been far below expectations, both for the current and future dated allowances. These low prices can be explained only in part by the start-up process, the hesitancy of traders, or the discriminant type of auction. More important are the cost-saving decisions made by emitters in response to market incentives. Workshop discussions revealed how varied these cost-saving choices are, and how innovations are occurring in the application of previously known control technologies and measures. Utilities carrying out smart wholesale power shopping and clever load management among generating units with varying emission rates per kilowatt hour produced are among the operational maneuvers that lower emissions cheaply. Railroad deregulation has reduced transportation costs of low sulfur coal, broadening the market for that fuel. What is innovative is the new train arrangements that further reduce effective transportation rates, and the use and blending of low sulfur coal in ways formerly thought to be difficult, if not impossible. The unexpectedly low prices of allowances, an indicator of these lowered marginal control costs, are clear signals that the market is working in ways unknown to command-and-control.

Transaction costs appear to be lower than once anticipated. To all reports, brokers have facilitated trades and, in some cases, have custom-designed financial derivatives to hedge or shift risks. Early reports from the RECLAIM market for NO_x credits indicate that transaction costs are modest there also. Public acceptance is growing, and there is greater recognition that issuing permits to pollute is not an immoral but a regulatory action, designed to reduce pollution. Our conclusion is that cap-and-trade markets as a new incentive scheme are working surprisingly well.

The sulfur dioxide cap-and-trade market was developed in the light of the savings realized in the earlier emissions reduction credit programs and in the successful lead reduction program, among other incentive applications. This history is not one of untrammeled successes; rather, it is an account of progress in discovering the limitations of particular forms of existing incentive systems and in improving the design of new systems. Another ambitious design is to be found in the (partial) cap-and-trade market about to be launched to reduce stationary source volatile organic compound emissions in northeastern Illinois.

This Illinois plan has garnered widespread support and appears to be solidly in place for a start in 1999. Among the questions to be addressed are those raised by the nature of the pollutant. Sulfur dioxide is a complex molecule, contributing to acid deposition, particulate matter, and other harmful substances, but it is a relatively simple pollutant compared with the "hot-spot," varying potency, and vector or multidimensional character of ozone precursors like the hydrocarbons.

These complexities cannot be swept under the rug. The regulatory agency's present policy is to treat the hydrocarbons as a uniform and well-mixed pollutant, which facilitates a simplification of market rules and tradable emission allowance design, thus increasing the chances that traders will participate effectively. The discernible, if limited improvement of air quality brought about by cleaner cars and other CAC regulation was not achieved by any systemic treatment of ozone complexities either. The improved modeling and more detailed monitoring of regional air should provide future feedback information that would enable the regulating agency to change its present policy, if required.

The Workshop participants gave this market a better than fifty-fifty chance of success after a detailed review of its design. They also offered encouragement to future applications, some of which are on the horizon. The Ozone Transport Commission and agencies in the northeastern United States have made progress in developing markets for ozone precursor control that extend beyond the boundaries of nonattainment areas. A voluntary association of states (the Ozone Transport Assessment Group) covering a large part of the country east of the Rocky Mountains is working on a trading plan, among other control mechanisms, to limit the long-range movement of NO_x. The U.S. EPA is considering its own initiative for NO_x trading over larger regions.

Corrective taxes set at levels that can achieve our environmental goals have well-known strengths and limitations of their own and offer a decentralized alternative to markets. Typically, fees charged for pollution are levied for revenue and not for corrective purposes. There are few applications of the principle enunciated by the economist Pigou some 80 years ago—that social costs ought to be distinguished from private and that a tax can be levied to correct for the discrepancy. Estimating the appropriate tax rate has proved difficult, and obtaining the consent of the taxpayer has been an even more formidable obstacle. The editors join with those who propose green taxes for the control of pollutants and for the generation of revenues that can replace distorting taxes on work and savings. This "revenue neutrality" may make green taxes a more attractive option to the taxpayer.

A few final comments. Given our favorable conclusions about environmental market performance and prospects, what suggestions do we have for further applications? What unresolved matters require further basic and applied research? We end with a discussion of the possible influence of the Workshop setup on the studies of this book.

TOWARD MORE EXTENSIVE USE OF INCENTIVE SYSTEMS

We believe there are many good reasons for the heavy emphasis in this publication on environmental markets. Their attractive economic and political properties make no apology necessary, but there are other important market-based approaches that have not been covered in detail. We did venture, in Part 3, into new territory, such as green taxes and voluntary environmental quality control programs, but by no means do these two studies exhaust the list of additional problem areas amenable to more extensive use of incentive systems. Some examples are in order.

Water Quality Control

Water quality has remained the epitome of prescriptive environmental policy characterized by technology-based discharge limitations and the National Pollution Discharge Elimination System. Some flexibility would appear to be a possibility under the Clean Water Act of 1972; the waste load allocation between point pollution sources in nonattainment areas could, in principle, be shifted among sources to lower costs as long as the total maximum daily load is not violated. Drinking-water treatment costs have, on occasion, been lowered by communities paying non-point pollution sources to keep nutrients out of water supplies. Neither incentive idea has been sufficiently developed to yield significant savings.

Despite locational features, which figure prominently in water control designs, and the fact that variability of control costs among sources of water pollution may be less than among sources of air pollution, there would appear to be much room for reform of clean water legislation. "Smart" combinations of control measures could be developed, including revenue-raising taxes on polluting inputs, tradable effluent allowances, and green payments that encourage environmentally friendly agricultural practices in lieu of crop subsidies (Braden et al. 1994). Impediments to incentive applications for water quality control may be receding, and a continuity review of regulatory reform and revision of the underlying act would be valuable and timely.

Pricing of Municipal Waste Disposal

Disposal of municipal postconsumption solid waste has led to deeply held convictions among concerned groups about the appropriate waste disposal or mix of disposal measures. However, there has not been sufficient benefit-cost analysis to support, convincingly, one view against another. The costs of landfilling have changed as standards have been increased; incineration problems have multiplied; and waste prevention remains at a rudimentary stage. Obtaining information on recycling has not been easy. Most cities in the United States charge a fixed fee for waste collection service— usually buried in the general tax assessment—regardless of the amount of waste generated by individual taxpayers.

Raising the zero marginal cost for collection now prevailing in many communities by some form of unit pricing for nonrecyclable waste would introduce an incentive to prevent waste in product design and manufacture and to encourage recycling. The tax on nonrecyclable trash ought to be set where the extra cost of collecting a ton of trash equals the harms or social costs of that ton, admittedly a difficult calculation that may call for rough approximations. Allowing the market more rein in this respect, and more rein in determining the prices of recyclables, ought to help the way toward a more informed and cost-effective combination of waste prevention, incineration, landfilling, and recycling.

Other possible applications of incentive systems are not hard to find, many of them extending the boundaries of the uses of market-based approaches. Tradable property rights for water usage, for access to fishing areas, and for habitat trade-offs are only some of the applications receiving consideration. As more applications are tried and tested, the more valuable the information base will become. Adding to this information base has been one of the objectives of this volume. An earlier discussion of the general topic may be found in a current, readable, and broad survey by Tietenberg (1996).

RESEARCH DIRECTIONS

Continued and more intensive research into incentive systems was called for by our contributors and workshop participants. The deliberative opinion poll was one mine of suggestions from the latter group. The authors in many of their studies made detailed recommendations for further work.

In the case of the performance of the specific market being designed for northeastern Illinois, there are important questions about the location of emission sources — the "hot-spot" problem, the varying potency of different hydrocarbons in contributing to ozone formation, and the control of toxic hydrocarbons that remain on the research agenda. Most present market designs exclude these complications in an effort to get a workable scheme under way. That policy is defensible, but it is important to confirm it with additional evidence.

Effective inclusion of other sources of ozone precursor emissions, especially mobile on-road vehicles, in a tradable emission market has been high on everybody's agenda but, as yet, implementation of this idea has, so far, proved difficult. It is possible that prescriptive regulation remains the tool of choice in the case of numerous small sources of pollution. It is too early to close the book on incentive plans, however.

Emitter trading behavior in the market remains to be studied more closely. That variability of control costs exists among stationary sources is confirmed by studies of the Illinois EPA. The potential for savings generated by active trading of pollution rights is there. For effective functioning of the market, emission sources will have to manage their tradable permit portfolios in a manner to take into account the degrees of uncertainties in prices, regulations, and control technology. There will be new cost and trading calculations to be made.

The nature and extent of transaction costs, sometimes said to be the Achilles' heel of environmental markets, should attract more research attention. These costs have to do with search efforts to locate traders on the other side of the market. They have to do with efforts to discover present prices and to estimate future prices. They have to do with negotiation and bargaining efforts once the deal is on the table. The role of brokers and market makers in this connection deserves attention. Declining transaction costs have characterized many maturing markets, and it will be important to look for this trend in environmental markets.

These research issues carry over, in many instances, to the sulfur dioxide allowance market, but there are important differences. Electric utilities, under complex regulation already, may respond quite differently than other types of enterprises to cap-and-trade market opportunities. For example, the overhang of untraded allowances held by utilities raises the possibility that some utilities may not be choosing the least-cost option for meeting their emission requirements; that is, they may be reducing emissions with control techniques at a cost per ton above the going market for allowances. This apparently puzzling behavior invites research.

In the case of application of pollution taxes, to gauge the reduction in pollution, we need to know more about the taxpayer's response to various rates and willingness to accept pollution charges in exchange for a reduction of other, adversely distorting taxes.

The increased use of unit-based pricing for disposal of municipal solid waste programs will generate interesting cross-sectional data as municipalities of varying size and character try out their preferred arrangement. Studies of waste generator responses to pay-by-the-bag and other incentives could help sort out cost-effective means of disposal. A preliminary illustration of the value of such information may be found in Miranda et al. (1994).

The use of carbon taxes as a means of controlling emissions of carbon dioxide has frequently been modeled by economists, for example, Nordhaus (1993). However, these taxes have been much more studied than put into effect. A few countries have made tentative steps in this direction, but the evidence is neither sufficient nor persuasive. Perhaps no tax proposal, given the significance of the fossil fuel tax base and the magnitude of the tax rate that may ultimately be required, will generate as much opposition, and support, as the carbon tax.

This book deals with half of the environmental problem. Incorporating the other half, damages of pollution, into a more comprehensive cost-benefit analysis is brought to the fore by incentive plans such as cap-and-trade markets. Such additional research, when carried out with full recognition of its strengths and limitations, would contribute to evaluating whether the environmental targets we take as given are set at the optimal level.

Turning to the question of cost-effectiveness of alternative control measures, our opening study clearly indicated that data, while becoming more available and more valuable, continue to be far from complete. Savings in compliance costs can take several forms. Static cost-effectiveness results from simply reallocating emissions among emitters until their current marginal costs of control are equalized; that is, the emitters who can reduce most cheaply do so. An even more significant savings source may be found in dynamic cost-effectiveness or the development of new control procedures and equipment. The incentives are in place, but the evidence of their effect is as yet spotty and anecdotal. This, in our view, is a most urgent topic calling for more research.

CONTRIBUTIONS OF THE WORKSHOP

Having the papers on hand before each meeting, having ample time for discussion, and having the opportunity to deliberate on topics over a series of meetings, Workshop participants had many chances to raise informed questions, to challenge dubious assertions, and to express pointed views in both written and oral form. Recall that they were experienced members of regulated, regulating, and public interest communities. A considerable amount of contributor rethinking and rewriting did, in fact, take place under these conditions.

The exchange of views among participants about the performance and prospects of market-based incentives not only strengthened the editor's confidence in these measures but also helped clarify in our minds the most efficient roles for the concerned communities. Governments would continue to have the vital task of setting the quantitative goals of pollution reduction but not the detailed prescription of how to achieve those goals. The regulated community would then do what it does best: determine the most cost-effective use of control inputs at the enterprise level to reduce pollution. Environmentalists would do what they have done well in traditional regulation: call for more access to information, for more accountability, and for needed revisions in environmental policy in the public interest. Academics, finally, would do what they are well placed to do: estimate the benefits and costs of incentive schemes, and appraise the results.

REFERENCES

Braden, J. B., N. R. Netusil, and R. F. Kosobud. 1994. "Incentive-Based Nonpoint Source Pollution Abatement in a Reauthorized Clean Water Act." *Water Resources Bulletin* 30(5):781–791. October.

Miranda, M. L., J. W. Everett, D. Blume, and B. A. Roy Jr. 1994. "Market-Based Incentives and Residential Municipal Solid Waste." *Journal of Policy Analysis and Management* 13(4):681–698.

Nordhaus, William D. 1993. "Optimal Greenhouse Gas Reductions and Tax Policy in the 'Dice' Model." *American Economic Review, Papers and Proceedings* 83(2):313–317.

Tietenberg, T. 1996. *Environmental and Natural Resource Economics.* New York: HarperCollins.

U.S. Environmental Protection Agency. 1990. *Environmental Investments: The Costs of a Clean Environment; A Summary.* EPA 230-12-90-084. Washington, DC: U.S. Environmental Protection Agency. December.

4.2

❑

CONTRIBUTORS

❑ **Robert C. Anderson** is currently a consultant on energy and economic incentive systems. Previously, he was research manager for the American Petroleum Institute. He has published studies in the area of incentive mechanisms for environmental management. He holds a Ph.D. from Claremont University.

Larry S. Brodsky is president and chief operating officer of Orange and Rockland Utilities, Inc. At the time of writing his comment, Mr. Brodsky was senior vice president of Illinois Power Company, where he had responsibilities during his tenure for managing fossil and nuclear generating stations, system operations, planning, and natural gas business activities. He holds a B.S. degree from the University of Illinois at Urbana-Champaign.

Ronald L. Burke is the director of environmental health at the American Lung Association of Metropolitan Chicago (ALAMC). Mr. Burke has an M.S. in technology and human affairs (environmental policy emphasis) from Washington University in St. Louis and has worked for both the U.S. Environmental Protection Agency and the Illinois Environmental Protection Agency. Since joining ALAMC, he has focused on the timely and effective implementation of the Clean Air Act Amendments of 1990.

Alan P. Carlin is senior economist, Office of Policy, Planning and Evaluation, U.S. Environmental Protection Agency, Washington, DC. He has published research on costs of controlling pollution and on marginal cost pricing of airport runway capacity. He received a Ph.D. in economics from the Massachusetts Institute of Technology.

Christian J. Colton was involved in sulfur dioxide allowance broker activities with Cantor Fitzgerald Securities Corp. of New York when he wrote his comments for this volume. He has subsequently joined the staff of the Peco Energy enterprise.

Daniel J. Dudek is a senior economist with the Environmental Defense Fund in its New York office. He holds a Ph.D. in agricultural economics from the University of California, Davis. He has been active as a consultant in many applications of market-based approaches, including the U.S. acid rain allowance trading program for sulfur dioxide. He serves as a member of the Science Advisory Board of the U.S. Environmental Protection Agency and has published numerous articles in his areas of interest.

Brian K. Edwards is a senior economist with the Law and Economics Consulting Group. He was formerly a senior economist with RCF Consultants. Prior to that he was an economist with the Argonne National Laboratory. He received an M.A. and a Ph.D. from the University of California at San Diego.

Cynthia A. Faur is an associate in the Chicago office of Sonnenshein Nath & Rosenthal. She received her law degree from the University of Chicago Law School in 1993 and has been practicing in the area of environmental law since passing the bar examination. Her work advising clients includes the use of economic incentives, emission trading, and conformity analysis. She has experience in the trading and valuation of emission reduction credits under Titles I and IV of the Clean Air Act Amendments of 1990.

Stephen L. Gerritson has been executive director of the Lake Michigan Air Directors Consortium since its inception in 1990. From 1981 to 1990, he was a principal in a national consulting firm, Economics Research Associates. Mr. Gerritson was also a member of the U.S. Foreign Service, stationed at the American Embassy in Lusaka, Zambia. He holds a B.A. in political science from the University of Massachusetts and an M.P.A. from the Woodrow Wilson School, Princeton University.

Joseph Goffman is a senior attorney at the Environmental Defense Fund, where he specializes in the use of market-based mechanisms for environmental policy attainment. He is a cofounder and director of the Environmental Resources Trust. In prior work, he was an associate counsel to the Committee on Environmental and Public Works of the U.S. Senate, where he had responsibilities for the development and drafting of Title IV of the Clean Air Act Amendments of 1990. He has been an active participant in the design of Illinois' VOC reduction and trading program, the NO_x emissions budget programs of the Ozone Transport Commission, the Ozone Transport Assessment Group, and the Maryland Department of the Environment. He received undergraduate and law degrees from Yale University.

Edward A. Helme is executive director of the Center for Clean Air Policy, Washington, DC. Prior to his present position, he was staff director for the Committee on Energy and Environment and director of the Natural Resources Division, both units of the National Governors Association. He is the author of a number of publications in the energy and environment fields. He holds a masters of public policy degree from the University of California at Berkeley and received a B.A. degree from Haverford College in political science and psychology.

Frances H. Irwin is currently a fellow in the Technology and Environmental Program at the World Resources Institute in Washington, DC. At the time of writing her comment, she was director of Pollution Prevention at the World Wildlife Fund. Her interests include environmental policy pertaining to pollution prevention, integration, and chemical harms. She has recently coauthored a book on establishing a benchmark for reporting on chemicals.

Roger A. Kanerva is environmental policy advisor to the director of the Illinois Environmental Protection Agency, where he leads that agency in developing proposals for market-based approaches to state environmental policy. He was a cochair of the design team for the trading plan for stationary source hydrocarbon emissions in the Chicago nonattainment (ozone) area and principal author of the final design proposal. He is a frequent speaker on emissions trading. Prior to joining the IEPA, he was deputy director for the Maryland Water Resources Administration. He received a B.S., and an M.S. in watershed management, from the University of Arizona.

Howard Klee Jr. is manager of business development for Amoco Orient Oil Company. At the time of writing his comment, he was director of regulatory affairs for the Amoco Corporation. He was a coproject director of the Yorktown Refinery Project with Dr. Mahesh Podar of the U.S. EPA. He received a Sc.D. degree in chemical engineering from the Massachusetts Institute of Technology.

Richard F. Kosobud is a professor of economics at the University of Illinois at Chicago, where he specializes in environmental economics. He is codirector of the Workshop on Market-Based Approaches to Environmental Policy, which has been supported principally by the John D. and Catherine T. MacArthur Foundation. His research, currently in the area of market incentives and the economics of climate change, has been widely published and has been supported by, among other sources, the National Science Foundation and the National Institute for Child Health and Human Development.

Karl A. McDermott was reappointed to the Illinois Commerce Commission on May 23, 1996, having been a commissioner since April 1, 1992. Prior to joining the ICC, he was the president and chairman of the board of the Center for Regulatory Studies (CRS), a lecturer at Illinois State University, a research scientist with Argonne National Laboratory, and a senior research associate at the National Regulatory Research Institute. He has a B.A. in economics from Indiana University of Pennsylvania (1976), a M.A. in public utility economics from the University of Wyoming (1978), and a Ph.D. in economics from the University of Illinois at Urbana-Champaign (1990). He is the author of many publications and reports.

Albert M. McGartland is director, Division of Economy and Environment, Office of Policy, Planning and Evalution, U.S. Environmental Protection Agency. He has published many articles on the uses and costs of incentive approaches to policy. He received a Ph.D. in economics from the University of Maryland.

Richard M. Peck is an associate professor of economics at the University of Illinois at Chicago, where he pursues his interests in the theory of taxation, state and local public finance, and property rights. He received a B.A. degree from Oberlin College and a Ph.D. in economics from Princeton University. He has published in a variety of professional journals.

Paul Pieper is an associate professor of economics at the University of Illinois at Chicago. He received his Ph.D. in economics from Northwestern University in 1984. His research focuses on issues of economic measurement, particularly those pertaining to government deficits, investment, and national income accounting. He is a recipient of two National Science Foundation grants and has published articles in the *American Economic Review, Review of Economics and Statistics*, and other leading professional journals.

Mahesh K. Podar is director of policy and budget in the Office of Water in the U.S. Environmental Protection Agency. He was a coproject director of the Yorktown Refinery Project with Howard Klee of the Amoco Corporation. Prior to joining the U.S. EPA, he was at the U.S. Office of Management and Budget and the Department of Transportation. He has a bachelor's and master's degree from Southern Illinois University at Carbondale and a Ph.D. from Pennsylvania State University.

Kevin G. Quinn is an assistant professor of economics at St. Norbert College, De Pere, Wisconsin. His research includes studies of intercountry carbon dioxide emissions trading for climate change control, among other environmental problem areas. He has worked as a policy analyst at the Argonne National Laboratory. He received a B.S. from Loyola University in Chicago and an M.B.A. in economics and marketing and a Ph.D. in economics, both from the University of Illinois at Chicago.

Robert C. Repetto is vice president of the World Resources Institute and director of its Program in Economics and Population. He is a member of the U.S. EPA's Science Advisory Board and the National Research Council's Board on Sustainable Development. His publications include many research papers on the economics of sustainable development. Before joining WRI in 1983, Dr. Repetto was an associate professor in the School of Public Health at Harvard University. Previously, he was a member of the World Bank's resident mission in Indonesia, economic advisor to the planning and development board for the government of East Pakistan, staff economist for the Ford Foundation in New Delhi, India, and an economist for the Federal Reserve Bank of New York. Dr. Repetto received a Ph.D. in economics from Harvard University, a M.Sc. degree in mathematical economics and econometrics from the London School of Economics, and a B.A. from Harvard College, where he was a member of Phi Beta Kappa.

Kenneth J. Rose is a senior institute economist in the Electric and Gas Research Division of the National Regulatory Research Institute of Ohio State University. He has worked primarily on studies concerning electric utility regulation and has contributed to many reports, papers, articles, and books. He was a project leader and principal investigator of the Institute's Clean Air Act, environmental externalities, and electric industry restructuring research. Before coming to the NRRI, he worked on energy-related issues at Argonne National Laboratory. He received his B.S., M.A., and Ph.D. in economics from the University of Illinois at Chicago.

William G. Rosenberg was an assistant administrator for air and radiation at U.S. EPA from 1989 to 1993, where one of his roles was managing the agency's contribution to the 1990 Amendments of the Clean Air Act. He is a leading advocate of market-based initiatives, regulatory negotiations, and consensus building in the acid rain and reformulated gasoline programs particularly. He is now president of E³ Ventures Inc., which provides assistance to businesses in designing proactive environmental strategies. He has served as chairman, Michigan Public Service Commission; assistant administrator, Federal Energy Administration; and executive director, Michigan State Housing Development Authority. He holds an M.B.A. and a J.D. from Columbia University and a B.A. from Syracuse University.

Manik Roy is director of the Pollution Prevention Policy Staff in the Office of the EPA Administrator. He has worked in the area of pollution prevention for the Environmental Defense Fund, the Massachusetts Department of Environmental Protection, and EPA's Office of Solid Waste. He has a Ph.D. in public policy from Harvard University and an M.S. in environmental engineering and a B.S. in civil engineering, both from Stanford University.

Richard L. Sandor is chairman and chief executive officer, Centre Financial Products Limited, a risk management firm specializing in derivative market applications. He is currently second vice chairman of the Chicago Board of Trade. He has served on numerous committees and boards including those of the Chicago Mercantile Exchange, the Banking Research Center of Northwestern University, and the Columbia University Futures Center. He was named "Father of Financial Futures" by the CBOT and the City of Chicago. Among other activities, he is an expert advisor to the United Nations Commission on Trade and Development on tradable entitlements for carbon dioxide emission control. He received a Ph.D. in economics from the University of Minnesota.

Ronald E. Schmitt is a senior environmental specialist in the Department of Environment, Health and Safety of the Amoco Corporation. He joined the corporation after graduating with a degree in mechanical engineering from the University of Illinois at Urbana-Champaign. He held a range of positions at Amoco in management and planning prior to his present assignment.

George S. Tolley is a professor of economics at the University of Chicago, where he was director of the Center for Urban Studies. He was deputy assistant secretary and director of the Office of Tax Analysis, U.S. Treasury. His research interests cover urban policy and theory, economic growth and tax policy, and agricultural resources. He has published numerous studies and edits a journal covering natural resource and energy issues. He received his Ph.D. from the University of Chicago.

Sarah M. Wade is a senior associate with Hagler Bailly Consulting, Inc. Prior experience includes a research internship at the Environmental Defense Fund. She received bachelor's and master's degrees from Yale University.

Jennifer B. Weinberger is a policy analyst with the Economy and Environment Division, Office of Policy, Planning, and Evaluation, the U.S. Environmental Protection Agency. She received an M.A. in public policy at the University of Wisconsin, Madison. She was a contributing author to *Technology for A Sustainable Future* (1994), published by the National Science and Technology Council.

Jennifer M. Zimmerman is currently a financial analyst in the investment research department of a major investment bank. She was codirector of the Workshop on Market-Based Approaches to Environmental Policy. Ms. Zimmerman received her B.A. from the University of Illinois at Chicago. A member of Phi Beta Kappa, she was graduated in May of 1996 with University Honors and Highest Distinction in Economics.

4.3

❏

INDEX

343